W9-CCF-211

LIFTING THE FOG OF WAR

LIFTING THE FOG OF WAR

•

ADMIRAL BILL OWENS

WITH ED OFFLEY

FARRAR • STRAUS • GIROUX • NEW YORK

Farrar, Straus and Giroux
19 Union Square West, New York 10003

Distributed in Canada by Douglas & McIntyre Ltd.
Printed in the United States of America
Designed by Abby Kagan
First edition, 2000

Library of Congress Cataloging-in-Publication Data

Owens, William A., 1940–
Lifting the fog of war / William Owens with Edward Offley.
p. cm.
ISBN 0-374-18627-8 (alk. paper)
1. United States—Armed Forces. 2. United States—Military policy. 3. United States—Armed Forces—Reorganization. 4. Military art and science—Forecasting. I. Offley, Edward. II. Title.

UA23.094 2000
355'.00973'0112—dc21

99-053312

CONTENTS

ACKNOWLEDGMENTS

This book is dedicated, first of all, to the men and women of the U.S. military, past and present, who represent a major part of U.S. diplomacy, security, and influence around the world.

I give special thanks to my friend James Blaker, who helped in every way with this book; and I am greatly indebted to Ed Offley for his fine and spirited work in collaboration. Ed would like to recognize his wife, Karen Conrad, their daughters, Elaine and Andrea, and his colleagues at the *Seattle Post-Intelligencer.*

Most of all, this book is dedicated to Monika and Todd.

LIFTING THE FOG OF WAR

INTRODUCTION

At the dawn of the twenty-first century, the United States remains the world's sole superpower, and the U.S. armed forces are operating in critical hot spots from Korea to Kosovo and from the Adriatic Sea to the Arabian Gulf.

Ten decades after the naval victories at Manila Bay and Santiago de Cuba in 1898 that propelled the United States onto the world's stage, our Navy warships steam unopposed in every ocean. A half century after its warplanes waged strategic bombing campaigns against imperial Japan and the Third Reich, our Air Force can launch precision-guided weapons at the enemy, or transport troops and equipment nonstop from the United States to any spot on earth. A quarter century after the decision to end the military draft, the U.S. armed forces are manned by highly educated, technologically adept professionals. A decade after U.S. ground forces triumphed in the Iraqi desert, our Army and Marine Corps are deployed on the ground from the Korean demilitarized zone to Kuwait. And ten years after the collapse of the Soviet Union, our strategic nuclear arsenal is still an overwhelming deterrent to any nation contemplating attacks by weapons of mass destruction on American soil or, indeed, anywhere else.

The United States today spends more on defense than the NATO allies, Japan, and South Korea combined. Our nation's $281 billion defense budget in 1998 (compared with the NATO allies' total of

$202 billion) represented 34 percent of all world military spending, up from 30 percent in the previous decade.[1] The Pentagon budget dwarfed the military spending of Russia ($40–$64 billion) and China ($37 billion).

But the U.S. military is in serious trouble today.

The armed forces have been reduced in size by 40 percent in the last decade, but are still largely structured as they were during the Cold War. The unrelenting pace of overseas deployments, humanitarian missions, and other unscheduled operations is straining military units and personnel to the breaking point. Combat readiness—the measurement of a unit's ability to carry out its wartime mission—is declining throughout the military. The four combat services—the Army, the Navy, the Air Force, and the Marine Corps—are struggling with unanticipated personnel shortages owing to a sharp decline in first-term enlistments and an exodus of experienced, mid-level career specialists drawn to civilian life by the booming domestic economy. As a result, military units are being forced to operate with as much as 10 percent fewer personnel than are needed to fill their ranks. And the Pentagon faces a crisis in its attempt to modernize the force because it has insufficient funds to purchase the next generation of warships, combat aircraft, and other military hardware to replace equipment that will become obsolete within the next decade or two.

There is no contradiction in those two descriptions of the U.S. military today. We have a topflight force that is running on empty, performing admirably with a growing number of weapons systems—including tactical fighter aircraft, Navy warships, and military transport helicopters—that are twenty to twenty-five years old and are becoming obsolete. Our men and women are suffering burnout as they are deployed in one crisis after another because our political leaders cannot understand the stress their policies have created. This situation cannot go on for much longer without dire results.

Many consider the debate over the future of our military a complex technical matter that is far removed from the daily concerns of the average American citizen. But the exact opposite is true: For decades we have enjoyed economic prosperity, burgeoning global trade, and political stability with our major allies because of the U.S. military's success in deterring aggression and safeguarding our national interests here and overseas.

We have become so accustomed to the benefits of our military security that we fail to recognize how vulnerable we would be to aggression, instability, and terrorism without it. Yet that is the precise risk we now face.

Despite its ability to carry out difficult and complex missions *today*, the U.S. military confronts a crisis in the decade ahead that—if ignored—could threaten a general collapse of its ability to execute the requirements of the current—or any future—national military strategy.

An assessment of the effectiveness of the overall U.S. national strategy itself, from which our military strategy is derived, would require a book of its own. But it's relevant to state here that in my assessment the Clinton Administration since 1993 has squandered our diminishing military assets on too many conflicts that did not genuinely affect the national security of this nation. As one recent report noted, "Between 1960 and 1991, the United States Army conducted 10 'operational events.' In the past eight years, the Army has conducted 26 operational events—2½ times that number in ⅓ the time span."[2]

The failure to devise a coherent strategy to protect our national interests and the failure to provide a military force capable of carrying out the strategy is putting those interests at risk, jeopardizing the lives of American servicemen and women, and weakening U.S. international leadership in an era when political chaos and military danger are on the rise.

As one group of experts warned last year, the U.S. military is heading for a budget "train wreck"[3] by the end of the new decade. By some estimates an additional *$100 billion per year* will be needed for defense spending to fully modernize, operate, and sustain the military force level identified in the Pentagon's 1997 Quadrennial Defense Review.[4] Conventional wisdom would dictate that the United States has two alternatives. Either the next President and Congress can opt to divert additional hundreds of billions of dollars into the defense budget in the year 2001 and beyond—to purchase the new weapons that are needed; to finance current operations overseas; and to pay for the expensive salary, pension, and "quality of life" programs essential to attracting and retaining new volunteer recruits. Or they can opt to reduce our military capability through inaction and default, a course that could lead to either a reduced U.S. military presence overseas or—worse—a return to the "hollow" force of the 1970s, which appeared robust but was seriously deficient in combat capability.

Our misadventure in Iran just twenty years ago should warn us of the consequences of permitting our military strength to erode. Just five years after the fall of Saigon and the end of our disastrous intervention in Vietnam, the United States was suddenly confronted by a major crisis in Iran after the fall of the Shah and the seizure of the U.S. embassy and fifty-three American citizens by revolutionaries. We were forced to send our overworked aircraft carriers to the Indian Ocean for nine months at a time, and attempts to mount a top secret rescue mission involving Army commandos, Marine helicopter pilots, and Air Force aviators failed disastrously when several aircraft collided in the Iranian desert.

My experience as a career Navy officer who rose to the second ranking position on the Joint Chiefs of Staff and my post-military work as a business executive involved in the global telecommunications industry has led me to conclude that neither of these "logical" solutions is acceptable.

Unless a true international military crisis directly threatens the physical security of the United States homeland, it is unlikely that any administration or future Congress will support substantial hikes in defense spending. For all of the heated rhetoric in Congress and the White House over the adequacy of defense spending, one observer notes that neither the Clinton Administration nor the Republican Congress has substantially moved to support a major hike in spending. In fact, the overall difference between Clinton and the GOP on defense has never been greater than 2 percent of the Pentagon budget.[5]

And to unilaterally withdraw the U.S. military from crisis areas or strategic regions would be to abandon our world leadership position and destroy the constellation of treaty partnerships around the world that have promoted regional stability and protected our allies and our own national interests for the last half century.

Fortunately, there is a third way by which the United States can retain its dominant military position in spite of funding constraints.

It's called the Revolution in Military Affairs.

But first, a personal explanation.

I never planned to become a revolutionary—especially after thirty-four years in the Navy. In fact, when I graduated from the U.S. Naval

Academy in 1962, all I wanted was to join and serve in the Navy's elite nuclear submarine service, a component of the armed forces known for its strict adherence to the ironclad rules and regulations instituted by the late Admiral Hyman G. Rickover, the "father" of the nuclear Navy.

It's ironic that this undersea branch of the Navy was defined for history by a technological revolution—the nuclear reactor—which transformed naval warfare, set the stage for the movement of the nation's strategic nuclear arsenal to sea, and dominated the naval standoff between the United States and Soviet Union throughout the Cold War. Ironic because by the time I joined it, the submarine service itself had become a place where leadership was defined by one's ability to follow established policy, and where individual initiative was considered a character defect. One close observer described Rickover (but also could have been describing the submarine service he so totally dominated) as "yesterday's visionary, who is today's conservative, and tomorrow's reactionary."[6]

The military is inherently conservative, which is not totally a bad thing. Part of an individual's sense of *place* within military service comes from the knowledge that he or she is serving in an honorable profession and in a specific service or branch that has direct lineage to the great battles and victories of American history.

For example, Army Rangers and paratroopers could take pride in 1998 when *Saving Private Ryan* appeared in movie theaters nationwide. Members of the elite 2nd Ranger Battalion at Fort Lewis, Wash., could see the patch of their unit on the shoulder of Captain Miller (the lead character, played by Tom Hanks) in a re-creation of the invasion of Normandy by three Ranger companies at sector Dog Green at Omaha Beach on June 6, 1944. And the soldier Miller's unit was directed to save, Private Ryan, wore the same "Screaming Eagle" patch still worn today by 15,000 Army soldiers in the famed 101st Air Assault Division at Fort Campbell, Kentucky.

Those of us who chose to serve in the military have much to be proud of and to commemorate.

Unfortunately, many military people are incapable of distinguishing between pride and blind loyalty to their specific military service. This attitude often is formed at the earliest moments of a military career (for example, the Army–Navy game rivalry). I confess that in my earlier years as a submariner I felt the same sense of "my" unit's superiority as my colleagues did, and I would pass on the various put-downs such as, "There are only two kinds of ships: submarines and targets." It sounds innocuous: but often enough such rivalries become institutionalized in the interservice competition for budget funds, or when the Joint Chiefs of Staff have to decide which service will lead an emergency joint task force at the outbreak of a crisis.

As I rose through the ranks and became entrusted with more responsibilities, I still kept the sense of identity and professional loyalty, first, to the submarine service, second, to the U.S. Navy, and third, to my country as a military officer. But I was fortunate to have two unusual career opportunities that allowed me to see the military universe from a broader perspective than is possible through the lens of an attack periscope.

The first experience came eight years after my commissioning, in 1970, when as a lieutenant I was selected as a Chief of Naval Operations Scholar and sent to study politics, philosophy, and economics at Oxford University in England. Three decades later I can still clearly recall the fury on Admiral Rickover's face when he summoned me to his Washington, D.C., office upon learning of my selection.

"Okay, Owens, goddamnit you can go to Oxford but you are going to study physics!" he shouted. (He always shouted.)

"Sir, in all respect," I replied, sweat pooling down my back, "I'm going to Oxford to study politics, philosophy, and economics."

(On returning I received another Rickover blast: *"It just shows you how low my program has sunk that I'd be willing to have you back!"*)

My second opportunity came in 1980, when then Chief of Naval Operations Admiral Tom Hayward selected me and a small group of other post-command officers to form the first Strategic Studies Group at the Naval War College at Newport, Rhode Island. For an entire year we were free to study, brainstorm, and travel to conduct an independent assessment of contemporary issues facing the naval service, which at that time was suffering greatly from post-Vietnam budget cuts, personnel shortages, and poor morale.

By the time I reached the rank of rear admiral in 1989, the Navy had helped make me aware that successful leadership required moving beyond the narrow perspective of one's own military service to embrace the issues affecting the U.S. armed forces as a whole. And more important, I had come to realize that the senior leadership of the U.S. military had to come to grips with the ending of the Cold War and the complex challenges that historic transformation of the world would bring.

It was shortly after my appointment as commander of the U.S. 6th Fleet in the fall of 1990—just as the U.S. military and our allies were mobilizing a large force to protect Saudi Arabia following the Iraqi invasion of Kuwait—that I began to learn how determined many of my senior military colleagues were to ignore the profound changes taking place.

I was stunned one day to receive an intelligence briefing on Soviet naval operations in the Mediterranean Sea. Since the late 1960s the Soviet Navy had maintained a continuous presence of diesel- and nuclear-powered attack and cruise missile submarines in the region, operating out of their Black Sea and Northern Fleets. To counter that, the U.S. Navy had kept a minimum of four nuclear-powered attack submarines (as well as an aircraft carrier battle group) deployed in the Mediterranean, which meant that about seventeen of the Atlantic Submarine Force submarines were continuously in training, preparing to deploy, or actually on patrol in southern European waters.

What I was told that day was that the Soviet submarines had vanished from the Mediterranean. They were no longer sailing from their home ports through the Bosporus and Gibraltar Straits to challenge the U.S. 6th Fleet. And my intelligence experts told me they appeared to have vanished for good.

So I sent a message back to Navy headquarters in the Pentagon that we could afford to relax our three-decade-old posture of attack submarine deployments. To be prudent, I suggested reducing our nuclear attack submarine presence from four to two subs for the time being.

Within a day or two I received a ferocious telephone call from a three-star colleague in the Pentagon.

"How dare you," he accused me. *"If you don't support us, our opponents will take advantage and use this to cut the force!"*

I was saddened and disappointed by this knee-jerk reaction. A sudden, unilateral reduction of force on the front lines of the U.S.–Soviet standoff was seen not as an opportunity for peace and budget savings, but as a direct threat to the parochial interests of the U.S. Navy and its submarine service.

That episode only hinted at the intense opposition and resistance I would encounter as I attempted to reform the structure of the U.S. Navy, and, as later, I led a small group of colleagues on the Joint Chiefs of Staff in our attempt to lay the groundwork for the Revolution in Military Affairs.

What is this revolution?

Simply put, the Revolution in Military Affairs (or RMA) seeks to use new technology to transform the way in which military units can wage war. For the commercial information technology revolution that is transforming American and Western societies today has powerful military applications, including advanced computer systems, global communications networks, and land-, air-, and space-based surveillance.

As one distinguished U.S. Army general put it not long ago, the RMA promises to provide the answers to five eternal questions every battlefield commander must ask:

- What is my mission?
- What is the enemy doing?
- How can I keep the enemy from knowing what I am doing and lead him to believe I am doing something else?
- Where am I vulnerable and does the enemy know it?
- Where is the enemy vulnerable and how can I exploit this and win at least cost to my soldiers?[7]

Strategists and military theorists have long written about the importance of knowing the battlefield and one's enemy—and have acknowledged how hard it is to do so. Over two thousand years ago Chinese General Sun Tzu gave advice that still resonates today:

> One who knows the enemy and knows himself will not be endangered in a hundred engagements. One who does not know the

> enemy but knows himself will sometimes be victorious, sometimes meet with defeat. One who knows neither the enemy nor himself will invariably be defeated in every engagement.[8]

In the West, men also have long turned their thoughts to coping with war's uncertainty. Still, despite the best efforts of the military commander to understand and find his enemy, uncertainty lingered. Napoleon Bonaparte wrote in his *Military Maxims*:

> A general never knows anything with certainty, never sees his enemy clearly and never knows positively where he is. When armies meet, the least accident of the terrain, the smallest wood, hides a portion of the army. The most experienced eye cannot state whether he sees the entire enemy army or only three quarters of it. . . . The general never knows the field of battle on which he may operate. His understanding is that of inspiration; he has no positive information; data to reach a knowledge of localities are so contingent on events that almost nothing is learned by experience.[9]

After the Napoleonic Wars, in which giant armies of infantry and artillery fought battles in all forms of terrain from Egypt to the Iberian Peninsula, and from Belgium to Russia, military theorists sought to understand the revolutionary changes in warfare wrought by Napoleon, particularly his creation of highly mobile and independent infantry corps commands and his improvements in the system for supplying troops with food, ammunition, and other necessities. Perhaps the objectives we should have in mind when contemplating a military operation were best articulated three hundred years ago by Swiss strategist Antoine-Henri Jomini: "To throw by strategic movements the mass of an army . . . upon the decisive points of a theater of war, and also upon the communications of the enemy as much as possible without compromising one's own."[10] Jomini admitted that this principle was simpler to state than to bring about, writing that "it is easy to recommend throwing the mass of the forces upon the decisive points, but . . . the difficulty lies in recognizing those points."[11]

Carl von Clausewitz's classic *On War*, published in 1812, was broader in its recognition of the political character of war.[12] Clausewitz wrote: "War is the realm of uncertainty; three quarters of the factors on which action is based are wrapped in a fog of greater or lesser uncertainty."[13]

From the T'ai Kung's eleventh-century B.C. advice to the Chou dynasty kings to Clausewitz's nineteenth-century treatise, military theorists have written about what they saw as an unyielding dilemma—the need to understand the battlefield, but the "certainty" that one could never sufficiently do so. Military theorists have always had to resign themselves to the fact that the fog of war would always be there, always cloaking and hiding what was actually taking place when militaries clashed. So the quest to see the battlefield—to know the enemy and thus surprise him, or prevent him from surprising you—has conditioned the ways militaries have prepared for war and, of great significance, the characteristics of the military forces we build today.

From the dawn of organized warfare until the development of the radio and the airplane, for example, the dimensions of the battlefield, and the commander's ability to influence events there, were defined by the range of human eyesight, occasionally aided by semaphore flags, heliograph mirrors, smoke signals, or the ability of a scout on foot or on horseback to personally deliver a message. Quite simply, armies and navies usually bumped into each other and then fought. The military term *meeting engagement* has long signified the unplanned encounter of armies that augurs a decisive battle—regardless of either commander's intent or preference. The Battle of Gettysburg in 1863 is a good example. Confederate General Robert E. Lee sent the Army of Northern Virginia into Pennsylvania in an attempt to strike the Union army and directly threaten Washington, D.C.; the two armies were groping for one another's location when a Confederate infantry brigade collided with a Union cavalry force, triggering the fateful battle.

Technology has changed the way we do battle. Better communications expanded the dimensions within which land, sea, and air scouts could report "what they saw" beyond visual range. When telegraphs, telephones, and finally radios enabled sensors, generally human beings, to report what they saw at increasingly greater ranges, technology that expanded the ability to see began to appear on battlefields, beginning with the Crimean War in 1856 and the U.S. Civil War four years later. During World War I, primitive reconnaissance aircraft and observation balloons were used to locate enemy troops and direct artillery fire. World War II brought a massive expansion of technology used to locate and pinpoint the enemy, including radar, radio direction-finding equipment, and sonar used to locate submarines. The Cold War saw the technology of surveillance taken high above the earth with the top secret U-2 spy plane, and later a galaxy of unmanned satellites that used photography, electronic eavesdropping, light-sensing devices, and radar to study the earth and its battlefields. By the time of the Korean War in 1950, and continuing through our involvement in Vietnam, technology redefined battlefields as much larger "operational theaters," and commanders set up headquarters farther and farther from the front lines in order to deal with the profusion of information that flowed their way.

Our ability to "see" a battlefield has improved, sometimes rapidly. Contrast our inability to detect the Japanese Navy's attack on Pearl Harbor until the warplanes appeared overhead with our better (although still limited) ability to detect, identify, assess, and attack the Japanese fleet at the Battle of Midway six months later. During the 1980s when U.S. forces were still poised to repel a Soviet–Warsaw Pact invasion of Western Europe, we were reasonably confident that we could detect Soviet attack preparations days ahead of any planned assault. General H. Norman Schwarzkopf, the commander in chief of allied forces in Operation Desert Storm in 1991, was able to "see" what was occurring in an area of nearly 250,000 square miles with dramatically greater fidelity and accuracy than any other previous military commander in any previous conflict. (His Iraqi counterparts, meanwhile, were so blind that the allied coalition was able to move hundreds of thousands of soldiers in heavy armor units west from the Persian Gulf staging ports virtually unnoticed, and so to prepare for the "left hook" offensive sweep that won the war.

Still, as late as the Gulf War, the fog of war—and certainly the friction—continued to pervade military operations. While Schwarzkopf and our other commanders probably had the best "view" of their battlefield in history, their vision was still spotty, usually late, and hardly clear. The technology and military systems used for battlefield surveillance did not cover the entire area at all times and could not see through bad weather conditions or, later, smoke clouds from the massive oil-field fires deliberately set by Iraq. The coalition's inability to locate and destroy mobile Iraqi Scud missile launchers and the ability of several Iraqi Republican Guard divisions to elude the anvil-and-hammer trap of the allied ground campaign are but two examples of how the fog of war and the friction of combat still existed in Operation Desert Storm. Because U.S. and allied war planners still appreciated these limitations, we used time-tested means of compensating for them: a massive air, ground, and naval force. In short, we conducted the campaign against Iraq essentially as Napoleon, Ulysses S. Grant, and Dwight D. Eisenhower had conducted their earlier campaigns, relying on all the means experience had shown were necessary to cope with the pervasive fog of war found at Austerlitz, Shiloh, and Normandy. We won the Persian Gulf War largely because we had learned the traditional lessons of warfare so well. And our victory confirmed traditional assumptions about the impossibility of seeing a battlefield in its full strategic dimensions—soon enough, clearly enough, accurately enough—to make a difference. Never in history—not in Napoleon's time, and not in the Balkans today—has a military commander been granted an omniscient view of the battlefield in real time, by day and night, and in all weather conditions—as much of the battlefield and an enemy force to allow vital maneuver and devastating firepower to deliver the coup de grace in a single blow.

Today's technology promises to make that possible. In fact, I believe the technology that is available to the U.S. military today and now in development can revolutionize the way we conduct military operations. That technology can give us the ability to see a "battlefield" as large as Iraq or Korea—an area 200 miles on a side—with unprecedented fidelity, comprehension, and timeliness; by night or day, in any kind of weather, all the time. In a future conflict, that means an Army corps commander in his field headquarters will have instant access to a live, three-dimensional image of the entire battlefield displayed on a computer screen, an image generated by a network of sensors including satellites, unmanned aerial vehicles, reconnaissance aircraft, and special operations soldiers on the ground. The commander will know the precise location and activity of enemy units—even those attempting to cloak their movements by operating at night or in poor weather, or by hiding behind mountains or under trees. He will also have instant access to information about the U.S. military force and its movements, enabling him to direct nearly instantaneous air strikes, artillery fire, and infantry assaults, thwarting any attempt by the enemy to launch its own attack. And the same powerful computer networks that make this possible will also grant the U.S. commander the ability to streamline the historically cumbersome supply process, making the whole force more mobile and therefore less vulnerable to attack.

Most important, the general or admiral will be able to immediately relay his orders (and the information that supports them) to his subordinate commanders through a computer network that includes video teleconferencing. In turn, subordinate commanders will be able to alert their units, brief the combat leaders, and prepare for battle in a fraction of the time required even today.

Having been a career military officer for thirty-four years, I believe the computer revolution, if correctly applied, presents us with a unique opportunity to transform the U.S. military into a lethal, effective, and efficient armed force that will serve the United States in the twenty-first century. This is the American Revolution in Military Affairs.

This new revolution challenges the hoary dictums about the fog and friction of war, and all the tactics, operational concepts, and doctrine pertaining to them. What's more, it challenges the entire conventional wisdom about how our military should be organized and structured—in ways that will have a profound significance as we try to allocate scarce federal dollars to modernize the U.S. armed forces for the new century. That's why it's a revolution.

In military terms, the current Revolution in Military Affairs falls into three general concepts.

The first concept is called *battlespace awareness.* This is a senior commander's overall comprehension of the enemy, his own forces, the battlefield terrain, and any other factors that will influence the course of battle. It rests on advanced sensing and reporting technologies and includes both the platforms and sensors we associate with intelligence gathering, surveillance, and reconnaissance and the reporting systems that provide better awareness of our own forces, informing us of the flow of logistics, the location and activity of combat units, and even the status of civilian noncombatants. Other elements include awareness of weather conditions, landscape features that will affect the movement and safety of forces, and electromagnetic conditions such as enemy radio jamming.

The second concept is known by its military acronym *C⁴I*, which stands for *command, control, communications, computers, and intelligence.* It entails creating a system within the military organization that serves as the central nervous system for the commander and his combat units. This network of computers and software will allow the organization to gather a wide range of *data* from unmanned sensors, other military units, and even the U.S. intelligence community; will enable trained specialists to gather and sift through the raw data to create *information* for the commander and his staff that is vital to their leadership; will help the commander to convert information to *knowledge* about his enemy, his own strengths, and the optimum course of action; then will make it possible to *communicate* that awareness to the rest of his organization. These concepts are not new—they describe the decision-making process of every military leader from Genghis Khan to Norman Schwarzkopf—but the power and speed with which these emerging technologies can assist the future commander *are* new.

The third concept is best described as *precision force use.* This involves the array of sophisticated weapons available to a U.S. military commander, including some that are in the arsenal today, such as precision-guided gravity bombs, laser-guided artillery shells, and cruise missiles that use navigational satellites for pinpoint strikes; and others that are still on the drawing board, such as lasers to physically destroy enemy missiles, thermal-pulse weapons to wreck enemy transmitters and receivers, and even computer programs to target and destroy the information network of the enemy by planting computer viruses.

This is not the first time military forces have been transformed by technology. There have been other Revolutions in Military Affairs in this century and in centuries past, and the thing they share in common is how national leaders and military commanders properly exploited technology to change the course of history.

Historically, the invention of the stirrup, which enabled armed men to fight on horseback, constituted a Revolution in Military Affairs. So too were the invention of the longbow, the musket, the radio, aircraft, and nuclear propulsion—each of which radically changed the way in which wars were fought.

But the current Revolution in Military Affairs, which began in military and civilian laboratories in the 1980s and moved into the Pentagon in the mid-1990s, promises to transform the application of military force far more than the stirrup, the machine gun, or even the nuclear weapon ever did. During the last four years of my military career, as deputy chief of naval operations for navy program planning in 1992–94 and as vice chairman of the Joint Chiefs of Staff during 1994–96, I had the unique opportunity to direct the first attempts to implement the current Revolution in Military Affairs into, first, the U.S. Navy, and, second, the armed forces as a whole.

I had the unique opportunity in the last eight years of my military career to witness the state of the U.S. armed forces at a major turning point in history from a wide variety of perspectives, including the office of the secretary of defense; a naval fleet headquarters; the Navy's Pentagon headquarters; and finally, the office of the Joint Chiefs of Staff. I was given the opportunity and challenge to play a key role—first with the Navy and then with the military as a whole—in transforming the military to adapt to both a changing world political environment and to an information technology revolution. My experiences in those assignments and my current observations on what I see as the Pentagon's faltering effort to implement that critically needed transformation have compelled me to write this book.

There has been progress. The Navy has undertaken a series of fleet battle experiments to test and incorporate high technology so as to expand naval power on existing and future warships. The Army is well on its way toward converting its ground combat units to a "digitized" force equipped with computer information networks and "smart" weapons that greatly increase both its mobility and firepower. The Marine Corps in its Sea Dragon initiative and recent "Urban Warrior" field training exercises is using the promise of digital computers and new technology to improve its ability to carry out amphibious attacks and to fight in hostile cities, where the physical environment and dense population make it extremely difficult to operate. And the Air Force has launched a number of long-term studies on implementing RMA technologies into the service.

The global explosion in information technology that makes the Revolution in Military Affairs possible is transforming all other areas of human society, from economics to politics to basic social organization. The impact of technology on military operations and national security will not disappear simply because the U.S. military leadership refuses to acknowledge reality.

I have been sad to see resistance and outright bureaucratic opposition toward the RMA at the highest levels of the defense establishment. This reaction is understandable, given the inherent conservatism of U.S. military culture, which itself reflects the civilian community from which it comes. And it is a point worth exploring at some length here.

It's a very human desire to be cautious and suspicious of change, particularly rapid alterations of the status quo. For people working within a large organization with deeply rooted traditions, there are some very comforting things about continuity, and the effort to achieve it is embedded in the notion of culture.

Regardless of the particular society or time, we humans design the attributes of our cultures largely to buffer ourselves against change. Our concepts of law, education, and politics can all be understood as efforts to temper, control, and deny change. And because the concern about rapid change is so deeply embedded in the human character, change doesn't often come easily in human affairs. Niccolò Machiavelli's admonition in *The Prince* about the barriers to changing existing paradigms, written some 450 years ago, still sounds familiar:

> And one should bear in mind that there is nothing more difficult to execute, nor more dangerous to administer than to introduce a new system of things; for he who introduces it has all those who profit from the old system as his enemies, and he has only lukewarm allies in all those who might profit from the new system.[14]

Yet however much we might wish to maintain the status quo, the world is never static. And even if we hope otherwise, human organizations—from corporations to governments, from clubs to societies—change, sometimes precipitously. This effort to hold on to the present—to grasp tightly and sometimes blindly what *is* rather than seek what *could be*—is particularly characteristic of military institutions. Change in military affairs is almost always uncomfortable, for military affairs carry high stakes. The men and women who choose the profession of arms generally understand this, which is why they work so hard to learn and capture the "rules" of operations in doctrine, training, and education.

But continuity is not the only value in military affairs. Sometimes it is very important to implement widespread change to doctrine, training, education, and technology. Mindless change is, of course, not particularly helpful. Facing up to the inevitability of change and appreciating the benefits of change is often required if an institution, including the military, is to prevail in a new era. Here we have no choice.

And it is dangerous folly for American political and military leaders to dismiss the possibility that the United States may face a serious threat from other nations or subnational entities that may use elements of the Revolution in Military Affairs against us.

Revelations that China has obtained top secret U.S. nuclear weapons design information from the U.S. Energy Department's nuclear laboratories confirms that Beijing is planning to become a full nuclear-weapons power in the new century—a lot earlier than many experts predicted. The terrorists who attacked the World Trade Center in 1993 reportedly made a bomb containing chemical warfare agents (which fortunately were destroyed by the blast), and a Japanese terrorist cult developed its own stocks of Sarin nerve agent which it released in the Tokyo subway in 1995, killing dozens and injuring thousands of innocent civilians. And several crude attempts at computer hacking linked to Serbia during the NATO campaign in Yugoslavia last year suggest a way our enemies might attack the U.S. military's worldwide communications network in the future. After all, much of this new technology is being produced by the commercial market and not as traditionally in government research labs. The nature of the information revolution is such that information-based technology can be used to attack us from unexpected directions, striking nodes and intersections of our military structure and civil society that remain unguarded by the standard array of military defenses.

This particular attitude of resistance against the Revolution in Military Affairs I attribute in great part to the residual overconfidence in many sectors of the American military a decade after our victory in Operation Desert Storm. As victors, we made the classic error, learning the wrong lessons from our experience. We used victory to validate doctrine, tactics, and weapons that had prevailed against a particularly inept foe. We ignored the fact that nations from China to Serbia have been studying the conduct of that war for a decade solely in order to devise counterstrategies and tactics for the next time we attempt an Operation Desert Storm.[15]

Historians tend to focus on the reasons for success. Their fascination is with victory, and although a lot has been written about why losses occur, most historical tracts address the reasons that one side won. Winners, after all, write the histories. Yet it's interesting and important to think about military failures—particularly when we are flush with victory, as America's military has been for most of the 1990s. There seems to be a common historical thread running through many of the greatest military losses. Those losses in great part stem from an arrogance that begets ignorance—an ignorance of facts and developments that others are quicker to see and to use. How was it that the outnumbered English defeated the flower of the French nobility at Agincourt? French knights in armor fell to English archers and foot soldiers because the very armor that signified their social station as knights and seduced them into believing in their invulnerability made them slow and vulnerable. Why did Confederate General George Pickett's charge against the Union army at Gettysburg fall short? Perhaps because Confederate General Robert E. Lee failed to recognize that the range of rifled weapons and artillery would butcher Pickett's division as it attacked across a mile of open fields. Why did France fall to the German blitzkrieg in 1940? France and Germany had the same weapons at their disposal and the French had more of them. But the two antagonists had taken different lessons from World War I, and the Germans were the first to recognize the powerful new synergy in the interaction of tanks, airplanes, and radio communications. Why did the Japanese, who pioneered the use of naval air power launched from aircraft carriers, miss the strategic significance of submarines and find their war machine starving when American submarines decimated Japan's merchant fleet?

Professional historians point to multiple reasons and caution against making generalizations regarding military defeats. Yet the more I read the historians' detailed explanations, the more it seems to me that there are common causes for military disasters, and at the heart of them lie dangerous smugness, institutional constraints on innovation, and the tendency to avoid questioning conventional wisdom. And the side that is the most smug, the most convinced that its interpretation of the past is the best guide for the future, often turns out to be the loser in the next war.

Modern history presents us with many examples of how a nation possessing a seemingly superior military force found itself suddenly defeated by an upstart neighbor who, according to the conventional wisdom, constituted no threat at all.

Consider France in the year 1860. As strategic analyst Michael Vlahos has written:

> It was the greatest military power of its day. Always the innovator, its technologies of war were the newest; its command and control the best of the best. Its ability to mobilize national resources was matchless; its traditions the envy and admiration of all. It was the world's military superpower.
>
> But now peace was everywhere . . . so the great nation refocused, turning its military toward new roles. It built a mobile deterrent force: A rapidly deployable army that could be rushed to quell a regional contingency (or two!) and restore stability. It reshaped its military forces around a core, professional force that could be used flexibly to achieve . . . the preservation of a stable world system.[16]

What happened? Ten years after that milestone, France and Germany fought the Franco-Prussian War, a little-remembered clash between the two neighboring states that took place between July 15, 1870, and May 10, 1871. Despite numerical superiority, more combat experience, and an apparent lead in weapons technology, France was decisively defeated on the battlefield, saw its armies crushed by superior German tactics, was occupied by a foreign army, and suffered a loss in power and international credibility that it never recovered.

Why did France lose?

While most historians seem to paint the German victory as a product of superior generalship (this marked the first appearance of what would become the admired and dreaded "Prussian general staff"), Vlahos notes that Germany exploited a civilian industrial-era technology that was available to France but that France had declined to exploit: the civilian railroad.

"It was the Big Change in ordinary life, the life of people in society, that had changed war," Vlahos writes. "The new artifacts of daily life were the bringers of revolution in war. The railroad was the tool of Europe's transformation, the agent at the heart of industrial revolution.

"This is the point of comparison (between the United States in the 1990s and France in 1870)," Vlahos adds. "We are in the midst of an economic upheaval equivalent to the industrial revolution in its capacity to transform our lives. Like the industrial revolution, this metamorphosis will reach up to politics and to war and remake them as well."[17]

I believe the U.S. military as a whole has failed to realize the true promise of the Revolution in Military Affairs. The separate services have conducted individual projects and field experiments, but these are largely being considered piecemeal. That is a prescription for failure.

Fielding high technology alone will not provide the U.S. military with the combat power and strategic agility we will need in the wars and crises of the future. The revolution must transcend new weapons and information systems to include a thorough reexamination of basic size, force structure, roles, and missions of the services.

To maximize the positive impact of this potential leap in technological superiority, the Defense Department must be willing to confront a number of controversial realities that the RMA presents us:

- First, we must be prepared to cast a ruthless eye over the current force structure and the roles and missions of the four combat services. This top-to-bottom analysis, predicated upon changing the structure to best incorporate and apply new RMA technologies, will be instantly controversial. For the Revolution in Military Affairs not only will suggest new roles and missions and combat organizations; it will likely also identify service components and commands, weapons, and platforms that are no longer needed.
- Second, we will need to thoroughly reexamine the current chain of command that runs from the President through the secretary of defense to the regional unified commanders under the Pentagon's Unified Command Plan. Just as a shift in geopolitical issues warranted the creation of the U.S. Central Command to handle the crisis area of the Persian Gulf, and the rise of terrorism and violent insurgences among allies sparked the creation of a unified U.S.

Special Operations Command, the ongoing revolution in technology led the Pentagon to assign to the U.S. Space Command responsibility for defensive *and* offensive warfare in cyberspace.

- Third, within the existing military force structure, we need to conduct a thorough reexamination of the industrial-age hierarchy that defines the U.S. military. Patterned after the Napoleonic Grande Armée, our system of divisions, brigades, and battalions may well prove to be archaic and top-heavy given the revolution in communications and networking made possible by computer power. The computer age is the age of networks, "flattened" hierarchies, and initiative being taken at the lower ends of any complex organization. Computer power and advanced communications will dictate new terms of effectiveness for military commands no less profoundly than for business or other private organizations.
- Fourth, we must reaffirm and truly implement the concept of interservice cooperation, and undertake to implement true "joint" military organizations to harness their mutual strengths. Despite legal mandates dating from the mid-1980s, today's joint military organizational structure is rarely joint. Rather, the array of task forces and unified commands consist mostly of components from different services that are administered and trained separately. When brought together they still suffer from insufficient interoperability and poor coordination.
- Fifth, the Defense Department must act to remove from the individual military services the power to make decisions on research and procurement of major weapons and systems. These powers should be transferred to a new Joint Requirements Committee in which the services have representation but the decision-making power resides with the secretary of defense and the chairman of the Joint Chiefs of Staff. Despite mandates under current federal law, directives from the Joint Chiefs of Staff, and pure common sense, many officials from the individual military services still regularly fail to consider that an effective multiservice military operation must draw upon the resources and strengths of other services.
- Sixth, the true Revolution in Military Affairs requires a transformation of the U.S. military's scattergun approach to combat doctrine, strategy, and tactics. It means we must transform the separate

> military institutions that draft military doctrine, and consolidate internal education and training into a single entity that can articulate a truly common vision. It calls for a transformed military educational system that inculcates the promises and requirements of the RMA into both officer leadership development and individual soldier training. And it means articulating a new Joint Vision 2015 document that encompasses the implications of RMA to the U.S. military as an organic entity, not as a confederation of competing military service bureaucracies.
>
> In addition, the Pentagon should move to consolidate the four services' redundant logistics, medical, and C^3I (command, control, communications, and intelligence) functions into separate joint agencies, each under the administrative control of one of the services. Again, this step would prevent needless and wasteful duplication and bureaucratic rivalry.

Each of these themes, if translated into budget reallocations or proposed force structure change, will immediately come under fire from established interests within and outside of the Pentagon. But the next administration and leadership of the Department of Defense cannot afford to allow traditional-minded leaders, military subgroups, the American defense industry, or their allies in Congress to thwart the successful execution of the Revolution in Military Affairs.

The debate and discussion over how to carry out this transformation should not be limited to the E-ring of the Pentagon nor to the traditional defense community. After years of neglecting to address defense issues, elected leaders, opinion makers, and average citizens must begin thinking anew about the use of military force as an element of U.S. foreign policy. We must again put the Pentagon budget in the center of the annual federal spending debate. We must also reexamine and affirm our commitment to carrying out the issues of diplomacy and conflict that are the responsibility of the world's solitary superpower.

More important, we must look with open minds to the profound transformation of warfare itself that is taking place before our eyes. The purpose of this book is to present an argument that the United States, facing a world of complex threats that will challenge our interests and citizens' safety throughout the next century, has no alternative but to embrace the Revolution in Military Affairs in its fullest consequences. It is my hope that this book can help spark an informed debate on how we can rescue our armed forces from the coming resource crisis and transform them to carry out that vital duty in the century ahead.

In the following chapters I look at the argument for expanding and accelerating the current Revolution in Military Affairs.

Chapter 1 begins with an overview of the U.S. military today—a globally engaged yet exhausted superpower force that has been stretched near the breaking point by multiple crises and peacekeeping operations in the decade since Operation Desert Storm. This review of the size and strength of the armed services includes the unavoidable facts that portend a major dissolution of American military strength—and perhaps even a total collapse of our military capability—within the next ten to fifteen years as weapons and equipment financed during the Reagan-era military buildup two decades ago become obsolete. This is the foundation upon which the need for the Revolution in Military Affairs begins.

Chapter 2 places today's military in the context of the vast technological and political changes that are uprooting the international structure and many of the assumptions about the use of military force, foreign policy, and international politics that we have depended upon over the last half century. I include in this presentation a review of earlier revolutionary periods in military history, when new technology suddenly gave a nation (or several nations) the capability to transform the use of force in pursuit of national objectives. This chapter also will share my own experience as a senior naval officer in discovering the implications of the Revolution in Military Affairs and the obstacles that immediately appeared when I and like-minded officers attempted to make initial reforms inside the U.S. military.

Chapter 3 focuses on the technological foundation behind the Revolution in Military Affairs. While military technology alone does not a revolution make—it's the integration of new technology with new organizations, operational concepts, and military structures that combines to present the revolution as a whole—technology is the core of the current revolution. Here I describe the technological applications I believe are driving the transformation of the U.S. military, and try to explain how they and other aspects of information technology can generate the synergy that will expand by orders of magnitude our future military prowess.

Chapter 4 examines in detail why the Revolution in Military Affairs—despite its strong support from the Pentagon leadership and Congress five years ago—has been stalled by bureaucratic opposition within the military and defense industry.

Chapter 5 is an analysis of the air war the NATO forces fought against Serbia in the spring of 1999. How much of the technology at the heart of the Revolution in Military Affairs was available for use in the Balkans? How much of it was actually used? Here I hope to make clear that technology alone cannot remake the U.S. military; a thorough reorganization of military structure and leadership is equally necessary. Kosovo, as we will show, was neither a military victory nor a validation of the existing U.S. military structure; rather, it was an expensive and inefficient use of existing American military technology that only hastened the oncoming crisis of obsolescence in our military. Within a few months, Kosovo had become a cautionary tale—hinting of the actual military potential in the Revolution in Military Affairs but also confirming how the lack of correct strategy and the wrong military organizational structure can nullify the best technology.

And Chapter 6 provides some specific recommendations on what is needed now to fully implement the Revolution in Military Affairs, based on my assessment of the current situation and the pending crisis in military resources. At minimum, I hope this chapter sparks debate on the future of the U.S. military, for it is a debate that the American public urgently needs to hear.

CHAPTER 1

THE EXHAUSTED SUPERPOWER

Although the twentieth century was one of the most violent eras in history, we can at least look back on the last five decades as a period that—while fraught with the threat of superpower conflict—still enjoyed an inherent stability because of mutual deterrence between the United States and the Soviet Union. In contrast, in the twenty-first century we most likely will face increased political turmoil, economic instability and chaos across a wider range of known and unknown threats ranging from hostile nations to terrorists armed with weapons of mass destruction. Despite the complex variables and unknowns, one fundamental truth exists: Given the current military strength of the United States as compared with that of other nations, and recognizing the extent of existing diplomatic, political, economic, and cultural ties between the United States and the rest of the developed countries, it is impossible for this nation to consider abandoning its position as a world leader. Central to that premise is the importance of U.S. military power as a pillar of our leadership.

But we are in jeopardy of losing our military edge within the next ten or fifteen years if we fail to transform the U.S. armed forces so as to incorporate the next generation of military technology, which will enable us to meet those new threats.

If we do nothing, the U.S. military itself faces a crisis within the next decade that could jeopardize our ability to protect American interests abroad and our national security at home. The crisis stems from three general factors: first, a decade of inadequate funding for daily military operations and unplanned missions such as those in Bosnia and Somalia; second, years of White House and congressional delay in funding research, development, and procurement of the next generation of ships, aircraft, and vehicles; and third, the failure of our political leaders to allow the military to implement meaningful cost-cutting steps such as the closing of obsolete bases and to refrain from adding meaningless pork-barrel items to the defense budget. Open-ended U.S. military commitments in Eastern Europe, Asia, and the Middle East especially are stretching the smaller, post–Cold War American military toward the breaking point.

The focus of this chapter is on the form and structure of the U.S. military today—the four armed services, the active and reserve components, the nine unified combatant commands,[1] the Pentagon leadership (the office of the secretary of defense, the Joint Chiefs of Staff, and the Joint Staff—the military headquarters organization under the chairman and responsible for planning and executing major military operations). It is a structure deeply rooted in the past, designed for a form of warfare we cannot afford today—based on massive operations using brute force to attain military objectives, and orchestrated by a military structure that is the product of five decades of wasteful interservice rivalry over budgets and bureaucratic turf. The current form of the U.S. military is a blueprint for military collapse.

Any debate over the future of military force structure should assume that the President (whatever his party), Congress, and the American people are unlikely to approve any substantial increases in defense spending above the current annual amount of about $250–260 billion.[2] The military funding realities underscore the critical need to redesign the military.

For most of the last century, military power has been a pillar of U.S. foreign policy, and there is no indication that this will change in the new century. In a report published last September 15 on trends in the international security environment, the United States Commission on National Security/Twenty-first Century outlined a broad spectrum of complex new threats and the severe challenges we will inevitably face in the decades ahead.[3] The panel, cochaired by former Senators Gary Hart and Warren B. Rudman, warned that the United States and other mature democracies around the world will be forced to take "concerted action" in the decades ahead to meet a host of threats unseen in the past fifty years:

- The United States will become "increasingly vulnerable" to hostile attacks on the homeland, and our current military superiority itself will not be enough to entirely prevent them.
- The ongoing revolutions in information technology and biotechnology will create new vulnerabilities for U.S. security as terrorists or hostile states seek to use those innovations for "nontraditional" attacks such as hacking information systems or launching biological weapons.
- The emerging global economy "will divide the world as well as draw it together," with the creation of a "transnational cyberclass" of highly educated people on the one hand, and a larger group of poorer people cut off from the Internet revolution on the other hand.
- The demand for fossil fuel—particularly oil from the Persian Gulf—will expand as the economies of major developing nations grow, underscoring the geopolitical stakes of areas with energy deposits.
- Although wars between states will continue to occur, there will be a proliferation of internal conflicts characterized by atrocities and the deliberate terrorizing of civilian populations (such as we have already seen in the Balkans, East Timor, and Chechnya).
- Space will become a critical and competitive military environment, and nations with space-launch capabilities will "likely" deploy weapons into orbit.
- U.S. intelligence collecting will become more of a challenge despite technological gains as subnational groups hostile to the United States and its allies both expand in number and become sophisticated in cloaking their activities.
- The United States will see the weakening of traditional alliances and will face pressures to reduce its forward military presence. Relying on ad hoc coalitions will make it much harder for the United States to find partners in response to regional crises.

The current role of U.S. military power under the National Military Strategy, most recently restated in the Pentagon's 1997 Quadrennial Defense Review, falls into three broad categories.

1. *Security.* The armed forces will continue to influence the international security environment by deterring hostile states and groups from attacking the United States or its allies, and by assisting friendly states while encouraging allies to shoulder a larger share of the mutual defense burden.
2. *Peacekeeping.* The U.S. military will continue to respond to a wide variety of military threats, ranging from major regional wars to small-scale peacekeeping efforts, humanitarian missions, or counterterrorist operations. The armed services must specifically prepare to face a threat of "asymmetric" warfare in which less powerful opponents use strategies, tactics, and weapons to offset or neutralize U.S. technological superiority—such as by firing chemical weapons at U.S. bases overseas to slow down or block entirely our ability to use them, or committing terrorist attacks against American civilians to distract us from military plans elsewhere.
3. *Global order and stability.* The United States must prepare for the dangers that do not exist today but may appear within the next twenty years, including the rise of a major military "peer" rival (China or a reconstituted Russia), widespread chaos stemming from the collapse of failing states in areas vital to U.S. national interests, or a wider proliferation of weapons of mass destruction.[4]

While the United States can take solace from the disappearance of the Soviet threat in the 1990s, the end of the Cold War has brought little in the way of worldwide stability. The United States and its allies in Western Europe and Asia have enjoyed overall political security, but the world as a whole has not. We have seen the rise of ethnic conflict and civil insurgencies (in the former Yugoslavia, Somalia, Indonesia, Rwanda, and East Timor), international criminal cartels (in Latin America and the former Soviet Union), nuclear weapons proliferation (in India, Pakistan, and North Korea), and massive natural disasters (in Bangladesh and Central America).

In nearly all of these instances, the United States has seen fit to intervene with military force—as a deterrent, as part of a rescue operation, as a peacekeeping force, or as a direct combatant. Experts warn that in the new century events such as weather pattern shifts and natural resource shortages could trigger political upheavals and mass civil unrest that could quickly escalate into genuine military crises.

As British defense journalist James Adams recently wrote, "Peacekeeping and warfare today are taking place in a world the likes of which we have never seen. All the old certainties have disappeared; there is no Cold War, no superpower rivalry to provide both tension and stability. Instead there has emerged a series of relatively small, unexpected crises."[5] But even in the smallest crisis in which we have recently intervened, the number of troops needed and the cost in dollars have risen higher than Pentagon planners ever anticipated.[6]

The broad goals of the U.S. national strategy generally reflect the political consensus that the United States will remain fully engaged throughout the world. But the response to the changing world political environment has been to give new roles and missions to our existing military force.

The threat of the spread of nuclear, chemical, and biological weapons—and ballistic missiles that can deliver them against the United States or its foreign military bases—has itself spawned three new Pentagon-led initiatives: a comprehensive "counterproliferation" effort to prepare U.S. military forces to intervene and seize nuclear, chemical, or biological (NBC) weapons from terrorists or rogue states preparing to use them; a massive training program (to be handed over to civilian federal agencies in the near future) to prepare civilian leaders in 120 U.S. cities to handle the consequences of an NBC attack by terrorists or rogue nations; and a broader, more general program to reorient U.S. forces for "homeland defense" against such threats.

During the 1990s we were blessed with a unique—but finite—window of opportunity to prepare the U.S. military for the future, owing to our nuclear and conventional military superiority at the end of the Cold War. But instead of using that time to reexamine and rebuild our military, we have in large part squandered it.

In retrospect, U.S. military force planning between the collapse of the Soviet Union and today is best understood as a cautious transition away from Cold War-era military requirements.

In 1990, as a junior rear admiral, I was serving as military assistant to Secretary of Defense Dick Cheney, a position I had held under his predecessor, Frank Carlucci. While I had close access to both the secretary and the chairman of the Joint Chiefs, my direct responsibility was not to formulate defense policy but to support my boss, the secretary, in his daily work and to act as a liaison with the senior military leaders in the Pentagon. What I saw during one crucial nine-month period—the time between the U.S. intervention in Panama in December 1989 and our decision to undertake Operation Desert Shield following the Iraqi invasion of Kuwait in August 1990—fueled my growing suspicions that the Pentagon's decision-making process was badly broken.

Cheney had reached past several dozen more senior four-star generals and admirals in the summer of 1989 to select Army General Colin Powell as chairman of the Joint Chiefs of Staff. As Bob Woodward later recounted in a book on the military, Powell's attractiveness as a candidate for the nation's highest military post stemmed from the variety of military and political positions he had held in the previous decade—senior military aide (the same job I held) to then Secretary of Defense Caspar Weinberger; national security advisor to President Reagan; commander of the Army V Corps headquarters in Germany; and, later, commander in chief of U.S. Forces Command, responsible for all Army units in the continental United States.[7] No one who served with Powell as I did can question the leadership, charisma, and moral strengths of this remarkable American, who rose from humble circumstances to become the nation's senior military commander.

But even Powell could not challenge the long-entrenched separatism of the four armed services, which paralyzed the decision-making process.[8]

By 1990, the senior military leadership in the Pentagon recognized that the Cold War was ending and asked, How should the armed forces prepare for this new era? I recall sitting in on several private discussions between Cheney and Powell on this issue as they struggled to formulate what would become the first post–Cold War restructuring of the U.S. military, called the *Base Force.*

This plan stemmed from Pentagon assessments in the late 1980s that the Soviet Union was facing severe economic strains that threatened its military strength, particularly Moscow's ability to sustain its massed air and ground forces in Eastern Europe alongside the Warsaw Pact coalition—assessments shown to be correct later that year as the Berlin Wall finally came down. While it was becoming obvious that the assumptions girding U.S. Cold War defense planning for so long were becoming obsolete, the Bush Administration and Pentagon leaders continued to direct the Defense Department to organize the military budget and long-range planning against the possibility of a "reconstituted" Soviet threat.

This hedge—that the old Soviet Union could easily recover to threaten NATO and the West—led Powell and Pentagon planners to temporize. The United States would reduce the size of its military without changing its fundamental structure. Thus each of the military services would take equal budget reductions of about 24 percent over the next five years. The resulting Base Force was intended to be a foundation from which we could as a nation rapidly expand our military capabilities to cope with a resurrected Soviet superpower—a scenario that became more and more unlikely as communism collapsed and the former Soviet military rotted in front of our eyes.

Powell, by nature a political animal, knew that the issue of deep budget cuts would strain his relationship with the heads of the armed services who served with him on the Joint Chiefs of Staff. He and I—and anyone else who worked in a leadership position at the Defense Department—knew that the Pentagon had *no* formal decision-making process through which he could even undertake a discussion of military changes that would hit one service harder than the others. I observed throughout that time that Powell had done a lot of thinking about the issue of the post–Cold War military but he kept close counsel only with his immediate staff. Powell knew there would be a real fight if the Pentagon leadership were to undertake to cut the services disproportionately—even if that was the right thing to do. The entrenched bureaucracy in the Pentagon, combined with Powell's assessment that genuine reform was a political impossibility, precluded any meaningful attempt to change the military. Several times the question came up about whether it might be possible to reduce "Cold War systems" (such as strategic nuclear weapons) disproportionately, and on several occasions I heard both Powell and Cheney acknowledge the need for more cuts in the force to free up much-needed dollars for new weaponry. But Powell's position hardened into insisting on the imposition of equal cuts across the board as the only way to reach a compromise.

Even had there been a willingness among our military leaders to debate openly about the force of the future, there was no way to really analyze what changes were necessary, no formal process to review possible options, and no objective process to approach such a momentous decision.

Six years earlier I had read with dismay defense analyst Edward Luttwak's scathing criticism of U.S. military mismanagement following a string of failed military operations in the late 1970s and early 1980s, including the Iran rescue raid, the failed 1982–83 Beirut peacekeeping mission, and the invasion of Grenada. He wrote bluntly:

> The failures in war and the continuing failure of peacetime have the same source. Shaped by laws, regulations, and military priorities that date back to 1945–48, the very structure of the armed forces and of the Defense Department are badly outmoded: and we now know that the system is quite incapable of self-reform.[9]

At the time I read those words, I did not want to believe them. But watching how the Base Force came into being with little study, no rigorous review, and a lack of forceful debate, I realized that the system itself was our biggest and most intractable adversary.[10]

Powell and Cheney, in their focus on appeasing the military services in the short term, set a precedent that would continue throughout the 1990s.

Entering office in January 1993, the new Clinton Administration immediately extended the delay of meaningful reform. Former Representative Les Aspin, the new secretary of defense, announced the formation of the *Bottom-Up Review* of the armed services, and the Pentagon staff duly churned away in search of a new set of post–Cold War planning assumptions. The Bottom-Up Review explicitly rejected the notion that the United States should design a military force meant to match a "reconstituted" Soviet superpower threat. But having proclaimed a new planning era, the review sketched out a force that was essentially identical to the one proposed in the Bush-Powell Base Force—only smaller. This tack reflected a Clinton campaign pledge during the 1992 presidential election that he would cut the U.S. military by an additional 200,000 troops on top of the 500,000-person reduction mandated by the Base Force.

The Bottom-Up Review postulated that the Pentagon should organize and structure the military to be able to fight two "nearly" simultaneous major regional conflicts in different regions of the world. Although the formula on the surface seemed to acknowledge the real-world situation, it was only a planning artifact—a means of keeping the military at its current size, as a recent congressionally appointed panel examining future U.S. military requirements concluded.

Some of the plan's provisions were sound: It emphasized the need for robust strategic mobility, something that most saw as very important as U.S. forces withdrew from bases overseas. It justified the heavy armored forces the Army had built over the preceding two decades. It supported a large naval aircraft carrier force for rapid responses, and a powerful Air Force able to provide rapid global response and overcome formidable air defenses.

By the time Aspin launched the Bottom-Up Review, I had returned to the Pentagon from command of the 6th Fleet and was directing a reorganization of the Navy headquarters staff under the leadership of Admiral Frank Kelso, the chief of naval operations (see Chapter 4). I was therefore the Navy's representative on an in-house Pentagon review team headed by Undersecretary of Defense John Deutch and cochaired by Admiral David Jeremiah, then the vice chairman of the Joint Chiefs of Staff. Again, as our meetings went on over a period of months, I was frustrated to see the opportunity for significant reform erode into a thinly veiled scramble among the services to protect their "share" of the overall Pentagon budget. There was still no formal process on the books by which the civilian leadership in the secretary's office could devise or evaluate the real choices. The military services, despite their own internal rivalries that dominated the halls of the Pentagon, held firm in their control over the debate. Meanwhile, there were voices in Congress demanding real change. The first call for an in-depth review of U.S. military force structure came in 1992 from the Senate Armed Services Committee, chaired by a true defense expert, Senator Sam Nunn, a Democrat from Georgia. Nunn proposed a Blue Ribbon Commission on Roles and Missions of the Armed Forces, staffed by independent experts to examine the existing system and particularly to determine whether there was an excess of overlapping military capabilities; for instance, all four services maintained combat air units. The Clinton Administration established the Commission on Roles and Missions in early 1994 under the chairmanship of John White. The White Commission produced a report eight months later, but by that time the 1994 congressional elections had resulted in a shift of control of the Senate and House of Representatives from the Democratic to the Republican Party, and Aspin had resigned as secretary of defense as a result of his decision to take full responsibility for the fierce battle that erupted in Somalia on October 3, 1993, during which 18 Army special operations soldiers were killed and dozens more were wounded in a vicious firefight with Somali fighters. (Somalia has become a textbook lesson in how a military can succeed on the battlefield yet lose overall. The special forces troops attained their initial objective in capturing several high-level aides to Somali clan leader Mohammed Farah Aidid on that day, and fought magnificently despite being outnumbered by several hundred to one. But the shock of the battle and high casualties exposed the U.S. policy failure that had allowed a nonpartisan peacekeeping mission to evolve into a low-scale war in which American troops had joined one of the sides.)

Back in Washington, the new principals—incoming Secretary of Defense William Perry and Deputy Secretary of Defense John White (the chairman of the Commission on Roles and Missions), as well as the new congressional defense committee chairmen—were understandably preoccupied with the tasks of assuming office and getting "up to speed" on the more pressing daily issues affecting the Pentagon.

Despite its formal title, the White Commission did not address the larger issues of service roles and missions. Instead, its report focused on two issues: the costs of the U.S. defense infrastructure (bases, research laboratories, and supply depots) in the United States and the defense acquisition process. The commission did, however, recommend that a Quadrennial Defense Review be undertaken by each new administration.

(Perry was not oblivious to the need for a major overhaul of the post–Cold War military. In 1994, at the recommendation of John Deutch and General John Shalikashvili, he brought me into the top leadership of the armed forces by securing my appointment as vice chairman of the Joint Chiefs of Staff, and firmly supported my efforts to move the military beyond its traditional service-oriented approaches toward buying weapons and fighting wars. This experience is recounted in detail in Chapter 4.)

Not until the summer of 1996, six years after the release of the Cheney-Powell Base Force analysis, did Perry's office begin preparing for the first quadrennial review. By the time the findings were released the following May, Perry had been replaced by William S. Cohen, the Clinton Administration's third defense secretary in three years.

I had retired in February 1996, but remained a keen observer of the military reform effort. I considered the initial quadrennial review to be a major disappointment. Although it paid lip service to the need for a transformation of the armed services to reflect the Revolution in Military Affairs, the review again extended the premise that all services would receive the same share of cuts regardless of changed world conditions. The Quadrennial Defense Review proposed some additional minor force reductions (again, spread evenly across the four services to minimize the political and bureaucratic pain) without focusing on the possibility of abolishing unneeded military weapons or units. The review did recommend future increases in Pentagon spending to fund purchases of next-generation weapons such as tactical fighters and warships in recognition of the serious lag in modernization programs.

The resource crisis confronting the U.S. military in the decade ahead will not stem from the conclusions of the Quadrennial Defense Review, however. It will be the result of the gap between the cost of implementing the recommendations of that review and the actual funding levels proposed by the Clinton Administration and approved by Congress throughout the past decade. As former chairman of the Joint Chiefs

General Maxwell Taylor warned over forty years ago, the military budget "is far more than a compilation of dusty figures of interest only to fiscal experts. It should be a translation into dollars of the military strategy on which our future security will depend."[11] The Pentagon budgets throughout the 1990s have sacrificed future procurement requirements in order to pay for current operations such as Kuwait and South Korea, and unanticipated expenses from major peacekeeping missions in Somalia, Haiti, and Bosnia. The bill is now coming due and it is staggering. According to one respected analysis, during 1996–2000 the Pentagon has suffered an aggregate shortfall in defense spending—the gap between the actual costs and the amount sought by the White House and appropriated by Congress—of as much as $488 billion.[12]

Throughout all this, thoughtful observers on both sides of the Potomac River still longed for a deeper assessment of the military in the post–Cold War era. In passing the Defense Authorization Act of 1997, Congress directed the Clinton Administration to appoint an independent National Defense Panel responsible for preparing a roadmap for future military force structure apart from the official Pentagon assessment.

Chaired by BDM Corporation President Phil Odeen, the National Defense Panel released its report in December 1997—a turning point in the post–Cold War debate over U.S. military force structure. The report called on the nation to focus on what the security requirements of the new era will demand, rather than on defending the status quo of American force structure. More significantly, the panel called for a "transformation strategy" to bring the U.S. military into the modern era in great part by capitalizing on U.S. superiority in new information technologies, particularly our strengths in computers and various space satellite systems. Yet subsequent Defense Department budgets left the vast bulk of the military structure untouched while barely addressing significant reforms.

Since 1991 both the Bush and Clinton Administrations, with the tacit support of a Democratic-led Congress until 1994, and a Republican-led legislature since then, have cut military spending in real dollars again and again, no matter that the nation's role as international peacekeeper has broadened; and in response to the cuts, the leaders of the four service branches have shaved their budgets wherever they can—postponing new equipment upgrades, making do with old technology, and suffering serious personnel shortages—rather than deal with the fundamental change in military policy that is required of them. Since 1991 there has been a steady reduction in the size of the military forces built for the Cold War, taking place despite a desperate bureaucratic struggle by the individual services to halt the steady decline in annual Pentagon budgets from the Reagan-era high-water mark in 1985. Military service planners have sought, with considerable success, to reduce U.S. military forces without jeopardizing their Cold War-era organization, internal force ratios, doctrine, or mission allocation. It was a short-term bureaucratic victory that has sown the seeds for disaster.

A MILITARY STRETCHED THIN

Since 1989 the Bush and Clinton Administrations have shrunk the U.S. military from 18 to 10 active combat divisions; from 15 to 12 aircraft carrier battle groups; and from 24 to 12 active fighter wings. More than 800,000 active duty and reserve personnel slots have disappeared through programs that accelerated early retirements, reduced first-term enlistments, and offered a variety of incentives for excess mid-career people to leave. The force today is 40 percent smaller than it was in 1986, having shrunk from 2.3 million men and women to 1.4 million. Even so, the U.S. military still deploys and stations troops overseas at the levels established during the Cold War: 100,000 troops in the western Pacific, about 27,000 personnel in the Persian Gulf region, as well as another 100,000 personnel in Europe.

The result is a military that is stretched dangerously thin. A General Accounting Office study early in 1999 revealed that all but three of the Army's active divisions either were committed to ongoing peacekeeping in Bosnia, South Korea, or the Persian Gulf or were training for future deployments. Only three Army divisions—the 25th (light) Infantry Division, the 82nd Airborne Division, and the 101st Air Assault Divisions—were conducting routine training at military bases in the United States.[13] The seventy-eight-day U.S.-led NATO air war over Yugoslavia in the spring of 1999 (which is examined in detail in Chapter 5), stressed the U.S. Air Force even more, according to initial reports from a "lessons learned" task force: With more than 700 planes involved in the air campaign and flying most of some 34,000 individual combat missions, the Air Force used a greater percentage of its aircraft in Yugoslavia than it did either in Operation Desert Storm in 1991 or throughout the Vietnam War![14]

The emerging crisis over U.S. military resources at heart stems from a decade of delay in implementing overdue management reforms, cultural restructuring, and new technology in the military services as the budget declined. Thus the military has not been able to cover the actual costs of operations, maintenance, unanticipated missions, and procurement requirements generated by the Cold War-era military structure.

For most of the 1990s, the Bush and Clinton Administrations and Congress have invested about 3.1 percent of gross domestic product on defense, an average of $250–$270 billion annually. When the Clinton Administration released its proposed 2000 fiscal year Pentagon budget last winter, it appeared to have reversed its earlier trends. The document seemed to call for a $12 billion hike in spending for 2000, with a total increase of $112 billion for Pentagon programs between 2000 and 2005. Unfortunately, the credibility of the complicated funding package quickly collapsed on further review. An analysis by Representative Floyd Spence, chairman of the House Armed Services Committee, quickly found that the "increase" hinged on a number of questionable accounting steps and assumptions, including $2.5 billion in cuts of previously approved programs, a $3 billion cut in badly needed military construction projects, and unfounded assumptions that future Congresses would be willing to usurp legal spending caps on defense.[15]

Representative Spence estimated that the Clinton Pentagon budget for 2000 fell short by $150 billion over the five-year planning period 2000–2005 in terms of covering "critical requirements" identified by the Joint Chiefs of Staff. Other critics outside the Pentagon have charged that the gap is far worse, that Clinton underfunded the actual needs of the armed services by as much as $100 billion *per year*, particularly because of higher-than-anticipated operational costs such as peacekeeping missions in Bosnia, Haiti, and Somalia, the 1999 conflict with Yugoslavia, and a massive modernization requirement in the early twenty-first century to replace military weapons and systems first deployed in the late 1970s and 1980s that rapidly are becoming obsolete.

But another aspect of the problem is that the Defense Department has failed—despite the clear warnings from a decade and a half of declining defense appropriations—to shape a coherent decision-making process by which it can choose which systems to procure. The lack of a tough-minded leadership armed with a coherent process to make these hard choices threatens not only the short-term health of the armed forces but—more seriously—the long-term survival of the U.S. military as we know it, and the national security and foreign diplomacy that it makes possible.

One blatant example of the procurement dilemma is the current Pentagon plan for future tactical fighter programs. The Pentagon proposes developing three new fighter programs: the Navy F/A-18E/F Super Hornet, the Air Force F-22 Raptor, and the proposed Joint Strike Fighter whose multiple variants will be used by the Air Force, Navy, and Marine Corps. These three types of fighters will replace six entire types of military warplanes in the current U.S. military inventory over the next three decades. An almost unheard of movement within Congress last year to "solve" the problem by deferring future funding for the F-22 Raptor sparked panic within the Pentagon and defense industry, and supporters of the aircraft were waging a vigorous counterattack throughout the summer and fall of 1999 with the issue unresolved as this book went to press.

Some detail on this particular controversy will illuminate the magnitude of the resource crunch that is coming, and the Pentagon's inability to make the hard choices that are necessary.

Each of the three aircraft programs is designed to replace one or more existing fighter systems that have been in service with the U.S. armed forces since the 1970s. The F/A-18E/F Super Hornet is being developed to replace both the F-14 Tomcat and the early F/A-18C Hornets that now carry out the bombing missions and air-defense protection of the Navy aircraft carrier force. The Air Force has made the F-22 Raptor its top priority for new aircraft given the need to replace the service's stalwart F-15C Eagle interceptor, which will become obsolete in the next two decades. The F-22 Raptor will also carry out the important bombing missions now conducted by the F-117A stealth fighter by carrying precision-guided munitions in an internal bomb bay. And the Joint Strike Fighter (JSF), now scheduled to begin early production in 2005, is an aircraft "system" that will replace four different Air Force planes: the F-16 strike fighter, the F-117A stealth fighter, the A-10 Thunderbolt close air support fighter, and the F-15E Strike Fighter. The Navy plans to use a variant of the JSF for its carrier bombing missions, and the Marine Corps plans on a vertical-takeoff model to replace the AV-8B Harrier ground support fighter.

As currently proposed, these three aircraft will cost as much as $350 billion to develop and procure (assuming no major cost overruns)—more than 25 percent of the total estimated Pentagon procurement budget for the next three decades.[16] It is clear from the accumulation of news articles, government reports, and outside analyses on the subject that the proposal for procuring these three systems does not take into consideration existing budget realities or overall Pentagon requirements.

The Quadrennial Defense Review—supposedly the Defense Department's comprehensive analysis of future procurement—did nothing to set priorities with regard to the three aircraft programs, according to a 1999 study by the nonprofit Cato Institute that echoes the sentiments of other critics of the current Pentagon procurement system. The Cato report concluded that the anticipated defense budget crisis and the near-term absence of a major military threat to the United States justified canceling two of the three programs (the F/A-18E/F Super Hornet and the F-22 Raptor) while pressing ahead with the next-generation Joint Strike Fighter.

The Cato Institute study concluded: "The problem is that each service is putting its case for a new fighter in isolation rather than in the context of Department of Defense's overall fighter procurement program (it does not have one), much less DoD's overall procurement strategy for the next century (it does not have one, unless an unconstrained shopping list is a procurement strategy)."[18] And, *completely* absent in even a "perfect" fighter aircraft procurement process, is the question of whether the *balance* between aircraft and other elements of force such as ships, tanks, bombers, or RMA systems has been addressed at all!

It is my judgment that there is neither the economic means nor the political will in Congress or the White House to simply add $100 billion a year to the current defense budget, as outside experts advocate. What is equally clear is that maintaining the status quo in defense spending would mean embracing disarmament by default—at best, the country would return to the "hollow" forces of the 1970s, and, at worst, the U.S. military power would collapse within the next decade.[19]

There is a third course of action: By combining powerful computer technology and other modern information-based systems we could make a revitalized, leaner military force that is designed to outsee, outmaneuver, and outfight any foe. In this way, the United States can maintain and expand its military superiority within the budget constraints we anticipate over the course of the next two decades. But to do so means changing the U.S. military from a fighting force based on mass—overwhelming numbers of weapons and units—to a military centered on information-gathering technology applied to the next generation of precision weapons. And it means tearing down once and for all the fifty-year-old structure of the armed forces, which was built for waging war in the industrial age.

But before describing that revolution in technology and military organization, it's important to grasp the premises that have locked us into our obsolete military structure.

THE U.S. MILITARY TODAY

Because the United States is a continental landmass separated from Europe and Asia by the two oceans, the U.S. military is not shaped to defend U.S. borders against foreign invasion; it is an "expeditionary" force designed to do battle overseas. Unlike most nations, we have not had to fight on our own soil for over a hundred years, and since the "Indian Wars" in the late nineteenth century we have undertaken military campaigns only outside the continental United States.[20] Since World War II we have built and maintained a constellation of military bases in the Pacific, Western Europe and the Mediterranean, and (most recently) in the Middle East.

For the most part, U.S. forces deployed overseas bring what they need with them or have it supplied from the United States; they do not "live off the land" where they are stationed. A U.S. soldier and his family stationed in Germany or Japan is likely to live in housing built to U.S. tastes and specifications and perhaps from materials shipped from the United States. This military family will probably get most of its milk, meat, pasta, ketchup, beer, and potato chips at the U.S. base exchange or commissary. Their household appliances are likely to have been made in the United States, and the transformer needed to provide the correct voltage for their American appliances will usually come from America, too. If the family includes school-age children, chances are that their children will go to American schools run by the Department of Defense. Their military equipment is exclusively made in the U.S.A.; all spare parts come from the continental United States, as do many of the expendable, from medicines and whole blood to ammunition, cleaning fluids, and windshield wipers.

This pattern has historical roots. Most U.S. forces stationed overseas are in areas where their predecessors fought in World War II or took on occupation duties after the war. Those predecessors brought everything with them because little of what they needed was otherwise available, and the pattern continued in part because it worked. But we also stuck with this way of doing things because we never saw our overseas deployments as permanent. The goal of "containment" assumed that at some point the threat posed by Soviet communism would wither or collapse and U.S. forces would then return to the United States. True, as the years went by, Washington accepted that the confrontation with the Soviet Union would be protracted. Yet American policymakers never adopted an imperialist view and never saw U.S. military deployments overseas as something that would be permanent or that should become an integral part of American colonies abroad.

Over the years there has never been an equivalent to the underway replenishment systems the U.S. Navy has used to support its warships at sea, nor is there really any other nation so capable and adept at military airlift and sealift. No nation has developed global communications networks of the kind or the magnitude that U.S. military forces use. Few nations have similarly robust intelligence, medical, or administrative capabilities.

The organizational templates we used to build and maintain this infrastructure were the military services themselves. There were historical reasons for this; each of the military services had provided the channels, organizations, and procedures for supporting its own forces during World War II. Each force continued to do so as the focus of military planning shifted from war, to occupation, to deterrence and containment, and again to combat operations in Korea and Vietnam. A sailor looks for the Navy to pay, feed, move, and arm him, and to provide medical assistance when and if he needs it. He turns to Navy intelligence for information about the threats he may face and relies on the Navy to build and maintain the communications networks he needs to fulfill his missions. A soldier turns to the Army for these kinds of support. Airmen and Marines look toward their own services.

As a result, U.S. military forces are large—because they support expeditionary forces that can go anywhere in the world and fight there indefinitely—and there is a great deal of redundancy. This, in turn, affects what is called the "teeth-to-tail" ratio—that is, the combat portion of the force that seizes terrain, destroys targets, or tries to kill opposing forces, contrasted with the portion that supplies and otherwise supports these combat forces. The line between units that are clearly "teeth" or "tail" is occasionally blurred. But overall for every U.S. "combat" soldier or Marine, there are about 12 "support" personnel. About 15 men or women back up the combat actions of a single sailor. For every fighter-bomber pilot in the U.S. Air Force there are 32 other active-duty support personnel. Let me put it another way: for every combat soldier, sailor, airman, or Marine, there are up to three dozen more in the support infrastructure to feed, clothe, and arm him; transmit and receive his messages; fuel, arm, and repair his tank, ship, or airplane; tell him about the weather and the enemy; teach him; send him his mail; keep him entertained; and, should he die, retrieve his body, embalm it, and give it a proper burial.

Today 473,595 men and women in the U.S. Army work in positions that support the 60,000 troops that are specifically equipped and trained for actual combat. At the same time 363,449 active-duty Air Force personnel support the remaining 16,000 who actually fly or navigate combat aircraft or operate their weapons.[21]

And the support is not provided only by active-duty military personnel. In Operation Desert Storm in 1991, 30,000 civilian technicians, logisticians, and administrative personnel accompanied the 467,000 U.S. military personnel (including the activated reserve component personnel) deployed to the Persian Gulf area at the height of the buildup. Another 300,000 civilians—about a third of the directly and indirectly hired civilian employees of the Department of Defense—were directly involved outside the Persian Gulf area in support of the operation.

Some logistics rules of thumb may provide a sense of what drives these numbers. A soldier operating in the field needs about 26 pounds of food and water daily. During the Civil War, ammunition and horse fodder drove the average weight requirement per soldier per day to about 30 pounds. That figure rose to roughly 60 pounds per day by World War II. Currently, the average daily weight of food, water, ammunition, and fuel per soldier in a U.S. Army division is 400 pounds. Multiply that by, say, the 467,000 troops the United States deployed to Saudi Arabia in Operation Desert Storm and you get a rough idea of what it takes to support operating forces—over 93,000 tons of material each day.

All this overhead expense naturally has profound effects on the defense budget. Since well over two-thirds of the budget goes for the "tail," the efficiency with which we conduct these support functions can have a big effect on the costs of national security. This consideration often gets overlooked in the annual debates on whether the nation really needs to buy another B-2 bomber, a new nuclear attack submarine, or some other big-ticket item, because the support tail that goes with these weapons systems—which most of the defense budget is spent to maintain—is usually ignored.

How we apply technology to logistics, communications, intelligence, and medical support, then, has a direct effect not only on how we can fight but also on the visibility and political impact of our deployments overseas in peacetime. Powerful new computer-driven technology that can accurately tell us quickly and in detail the status of our forces—their inventories, their people, and the state of their equipment—can let us get the right kind of support to them when and where they need it. This information support really pays off in combat. It can mean "just-in-time" logistics, which will enable our combat forces to maneuver without the burden of a large supply train. It can mean faster medical support, and lower risk of loss of life in fights. But a better support mechanism can also mean a smaller "logistics footprint" in peacetime because information can substitute for the need for large storage areas. And smaller logistics footprints affect how prominent our forces are and how the surrounding populations see and interpret their presence.

To get a finer appreciation of how the current U.S. armed forces can be remade by the Revolution in Military Affairs, we have to look more closely at the U.S. military's people and equipment, its organization, and how it got to where it is now.

The Volunteer Force

Let's start with the people. With roughly 1.4 million people serving on active duty, the U.S. military is very large by current world standards. We have over three times as many people in uniform as Germany or France, for example, and nearly as many as all our NATO allies combined. Some other countries have more. China has roughly twice as many active-duty members of the military as we do. Russia, despite big cuts over the last decade, still has half again as many.

Almost half the Americans in uniform—as is true in other nations'

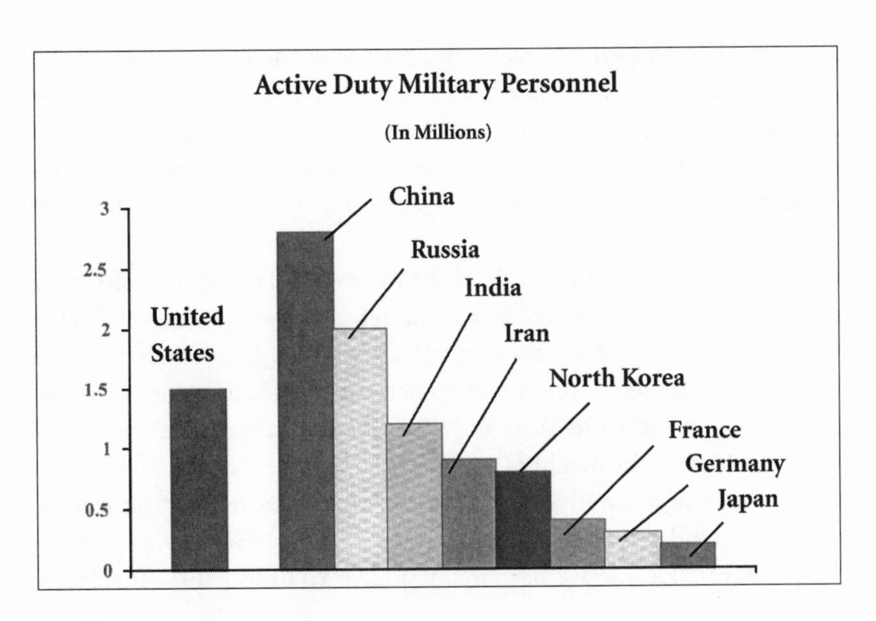

military forces—serve in the ground forces, that is, with the U.S. Army or the U.S. Marine Corps. But the number of people in the U.S. Navy and Air Force is strikingly high compared to other countries. There are 370,343 U.S. citizens serving in the Navy today, more than all the British, German, French, Japanese, Russian, and Chinese sailors combined. The 363,449 airmen in the U.S. Air Force, more than in any other air service, make up about 25 percent of the total number of U.S. active military personnel.[22] The average worldwide is 10 percent.

During the Civil War and World War I the number of Americans in uniform also placed the U.S. military among the world's largest forces. But after those conflicts the nation quickly cut its number of men in uniform. Only over the last half of this century has America maintained one of the world's largest militaries long enough for the size to seem "natural." Indeed, you have to be older than fifty to have lived in an America that did not have at least a million people under arms, and historically the size of our armed forces has embedded the military into the American society in unprecedented ways. The number of men and women serving in the military after World War II at any given time made up only between 1 and (at the height of our involvement in Vietnam) a little over 2 percent of the total population. Yet it was never precisely the same people. Because of the draft and the surge of veterans from World War II, Korea, and Vietnam, the percentage of the population who had had some direct experience with the military rose by the mid-1970s to almost 20 percent of the adult population, the highest proportion in American history. An awareness of—often pride in, sometimes anger toward—the U.S. military permeated American thought and politics deeper and longer throughout most of the last fifty years than it did ever before.

But the demographic profile of the military has changed profoundly since the end of the Vietnam War a quarter century ago. The active force of 1.4 million today is the smallest since 1949, the year before the Korean War. Today's force is over 30 percent smaller than the U.S. force at the end of the Cold War, and 60 percent smaller than the armed forces in 1969 at the height of the Vietnam War.

But more significantly, the proportion of the U.S. population serving in uniform today is smaller than in any year since 1930.

Moreover, because we have maintained a volunteer force for nearly three decades—a new arrangement that has dramatically changed the nature of both what types of people choose to join the military and how soon they get out—the percentage of the nation's population with any direct military experience has declined significantly. Today less than 10 percent of eighteen- to fifty-year-olds have served in the U.S. military, the lowest portion of this age group since before World War II. Ours is increasingly a smaller, professional military, and this development has weakened the military's traditional and important connection to the civil society it serves. Military service is a sociological norm in many nations; thirty years ago it was here, too. Today serving in the military is the exception to the rule for American society.

In 1997 the 186,000 new recruits that entered the U.S. military had all been born after the United States withdrew from Vietnam in 1973. And at the beginning of 1998 about 50 percent of the total U.S. officer corps—Army, Navy, Air Force, and Marines—had entered active duty after Operation Desert Storm in 1991 and have no direct personal experience or even firsthand knowledge of that conflict.

Less than 3 percent of the men and women on active duty today were in the military during the Vietnam War. All of today's Joint Chiefs of Staff had at least one tour of duty in Vietnam; most of the three- and four-star officers on active duty today did, too. But we were junior officers at the time of that conflict, so it is the post-Vietnam period that was the most formative for us; as we rose in rank and responsibility, we began to impress upon the military system our views of what the U.S. military should be and how it should operate. We sought to avoid the mistakes and blunders that we had lived through during the Vietnam War—most notably by organizing an all-volunteer force.

The people in today's all-volunteer military are older, better educated, more diverse, and better paid than their predecessors. Nearly half of all personnel are married today, up from one-third in the late 1970s. (As a result, those serving in the armed forces are a much more stable population, but they require many more family and "quality of life" programs.) Twenty years ago women held only about 15 percent of the active military's officer positions. Today they hold about 30 percent.

In 1975 the Defense Department began surveying young men and women about why they choose to join the U.S. military.[23] The surveys indicate that about 30 percent of the roughly 200,000 men and women who volunteered annually for active duty during 1996–98 did so to help finance their college education. (An eighteen-year-old who qualifies to join the U.S. military can accumulate a college education nest egg of up to $87,000 by the end of the first three-year enlistment term, including signing bonuses ranging from $12,000 to $20,000 for certain recruits, as a result of new recruiting initiatives announced late last year after the Army, Navy, and Air Force all failed to meet recruiting goals).[24] About 20 percent volunteered to obtain job training and experience, and another 20 percent joined because of the pay and travel opportunities. Only about 10 percent entered military service out of a sense of duty to the country, the survey found.

These figures suggest that those entering the military today see things differently than their predecessors of my generation, roughly half of whom volunteered out of a sense of duty.

Why such a profound shift? The world was a different place twenty years ago. President Jimmy Carter met Soviet General Secretary Leonid Brezhnev in Vienna in 1979 to sign the SALT II arms limitation treaty, but a few months later Soviet tanks rolled into Afghanistan, U.S. embassy personnel were taken hostage by Islamic revolutionaries in Iran, Iraq and Iran went to war, and the price of Gulf oil skyrocketed. The world suddenly seemed grimmer and more threatening than it had been. Beginning with that era of crisis following the post-Vietnam 1970s, the U.S. military would change considerably.

One of the differences with the all-volunteer force is that once in, a relatively large number of the volunteers tend to stay longer. True, quite a few people leave the military each year. To maintain the current level of active-duty enlisted personnel, the military must recruit about 190,000 people each year. Until very recently, enlisted personnel were staying in the military longer than their counterparts did a generation ago. In the 1990s, 80 percent of the new recruits stayed past their first term of service. In the 1970s, the figure was about ten points lower. Retention rates of officer personnel have always been higher than for enlisted men and women, but here, too, over the last decade, more officers have tried to stay in service longer than their predecessors did in the 1960s and 1970s. (One has to be careful in making generalizations about retention, of course. Retention rates for particular specialties can be much lower. Pilots in all the military services historically tend to leave earlier than other serving specialists—largely because of the demand for their skills in the private sector. Likewise, in periods of relative economic vigor such as we have now, retention efforts meet strong competition. Despite the draw of a robust economy, however, people in the military today still tend to stay longer than their predecessors. Higher pay and better benefits account somewhat for this pattern. Upward mobility and increasing responsibility count in decisions to stay in the military also.

If we look for the long-term significance of the all-volunteer force, we see two issues beneath the numbers, nuances, and trends. The first involves the people who choose *not* to serve in the U.S. military or who are excluded from serving. Today's higher recruitment standards mean that those who choose to go into the military are, on average, more educated than their predecessors and, on average, come from somewhat more affluent backgrounds. But while the uneducated and poorer are less represented in today's military, so too are the richer and more highly educated. The Pentagon tracking surveys indicate that over the last ten years, those who are most likely to go to college, most likely as a group to have higher incomes, and most likely to be politically active don't even think of joining the military.

This recruitment pattern hurts the military as an institution because the forces are less able to benefit from a cross section of American society. In turn, the nation does not benefit as it could from a cross section of the population having had direct experience with the military. By default, those who choose not to serve base their understanding and appreciation of military affairs, military institutions, and military processes on secondhand information rather than on any actual experience. This happens to be increasingly the case with the relatively highly educated, wealthy, and often politically powerful. In short, the people who are most likely to be the opinion leaders of the future are increasingly less likely to have a clue about military issues and may be prone to misuse the armed forces if they attain positions of national leadership.

The impact that an all-volunteer force could have on the fundamental constitutional issue of civilian control of the military is another issue that concerns us. Under current federal law, the President has control of the armed forces through the secretary of defense, who exercises authority over the chairman and the Joint Chiefs of Staff, and the regional military commanders in chief. Also the secretaries of the Army, Navy, and

Air Force are for administrative purposes the heads of the four armed services, although operational control rests with the defense secretary and his staff. Although the constitutional and legal principles regarding civilian control over the military remain sound, I believe actual control is becoming more difficult. The number of national political leaders who have had personal military experience has shrunk as an inevitable consequence of the end of the draft in 1973, and it is more likely that those who will assume civilian leadership posts over the military in the future may enter those demanding jobs with little knowledge of, or appreciation of, the day-to-day realities of military life.

Professionalism and Civil-Military Relationships

There is a positive side to the all-volunteer military. Because those who volunteer for military service tend to stay in longer—and over the last several years they have done so in the face of a lot of economic opportunity in the civilian arena—the military today is more professional and efficient than it was during the era of mandatory military service.

Military professionalism is synonymous with military efficiency. The more professional our military is, the better our military will be at doing the things it must, in the manner dictated by its doctrine and strategy. The U.S. military today is effective for the most part because the people filling the ranks understand how to fight, how their weapons work, and how to use force efficiently. Our military today can go into action sooner, more efficiently, and more effectively than it could in the past. Professional militaries are better at internal discipline. This is an important point, because the saddest events in American military history—the 1969 massacre of Vietnamese civilians at My Lai, for example—more often than not resulted from a breakdown of military professionalism rather than from an aberration in policy or regulations.

Like all institutions, the military spends a lot of time and effort on building internal bonds of trust, respect and sacrifice.

But while our military is very good at doing things as they are supposed to be done, it is not always good at changing the way things ought to be done. Highly professional militaries can be very good at maintaining the institution's traditions, mores, and cultures in the face of rapid and important change. This tendency can generate significant discrepancies between the military culture and that of the surrounding society, potentially causing grave misunderstandings—if not disturbing controversies. That happened in the Navy's Tailhook Association sexual assault scandal in 1991, for example, when the service's understandable celebration of its courage and power in Operation Desert Storm seven months earlier degenerated into a series of drunken encounters between male servicemen and female Navy officers and civilians, bringing great embarrassment to the service. When something like this happens, of course, it erodes the internal bonds of trust and respect as well as the trust and respect that must link the military to the civil society at large.

Professionalism within the military tends to generate an institutional conservatism that works against needed change inside the organization. This occurs because the military culture too often confuses *professionalism* with *loyalty* to a particular military service, or even to a professional specialty within a service (such as the Army infantry, naval carrier aviation, or Air Force fighter communities). The problem occurs when this relatively healthy expression of solidarity to a community hardens into an unreasoned, blind commitment to existing doctrine or structure.

Equating professionalism with automatically defending the status quo can be disastrous. This is the mind-set that drives service loyalties toward narrow parochialism, and congeals organizations into brittle shells. In this environment we end up ignoring new opportunities that could actually offer higher military effectiveness.

Finally, there is a political dimension to military professionalism that bears watching. Like any organization that identifies itself as distinct from the wider society, military people can be vulnerable to the elitist attitude that only they—and not the wider civilian community—know what is best in military matters. I believe that this attitude has become more widespread in the military over the past twenty-seven years with the end of the draft and the shift of the armed forces to an all-volunteer military, with the unintended consequence that the two communities have fewer connections than during the Cold War. (The role of the soldier in a democratic society attracted significant research in decades past but is not an area of much interest today.[25])

One notion we can dispense with is the Hollywood fixation on the military challenging the American civil authorities outright. Although we are regularly presented with motion picture plots, such as *Seven Days in May*, that portend a senior military commander usurping the President to take control of the nation, the reality is that American military leaders are professionals whose bedrock personal commitment begins with their commissioning oath *to serve and protect the Constitution of the United States*—not a political party or a particular President.[26]

On the contrary, a continuing gap between military leaders and civilian officials could well result in situations where civilian officials fail to press and to challenge the military leadership on serious issues. A good example of this is the controversy over Persian Gulf War veterans who were sickened during their service in Operation Desert Storm. It took nearly three years from the outbreak of medical symptoms in 1992 until the White House finally responded to Defense Department and Veterans Affairs Department inaction by appointing an outside presidential commission to study the matter.[27]

We should be concerned if the gap between military and civilian communities widens to the point that civilian leaders believe only the military can make competent decisions about military issues. This state of affairs could come to pass if civilian authorities do not care about or understand military doctrines. It can emerge from programming and budgeting processes that yield to military officials the decision-making power. It is more likely to happen when the military profession is highly respected, yet when public interest in military affairs slackens, and when the sons and daughters who make up the military ranks come from a narrower segment of the population. If you think this sounds a bit like a description of the United States today, you're correct.

Equipment

The United States military has more advanced weaponry than any other nation on earth.

The U.S. Navy operates a fleet of 12 aircraft carriers—all but a handful of which are nuclear powered—that can patrol in any ocean of the world. Our Navy fleet of 18 Trident ballistic missile submarines operating in the Atlantic and Pacific Oceans can threaten any hostile state with nuclear annihilation, thus ensuring that we will face no such threat.

Only the U.S. Air Force maintains a small but potent fleet of stealth combat aircraft, including the F-117A stealth fighter (which actually functions as a tactical bomber); the next-generation B-2A stealth bomber, which can strike targets nonstop from its base in Missouri and hit any point on earth; and the prototype F-22 Raptor, which—if it survives a congressional push to cancel funding—will combine stealth and high-technology precision-guided weapons.

The Army and Marine Corps today, although not the largest combined ground force in the world, are the most capable, with weapons and tactics honed to provide them with the ability to serve and fight in all kinds of terrain. From the M1A1 Abrams tank to the V-22 Osprey (the hybrid aircraft that can hover like a helicopter then transition into fixed-wing flight), the two ground services have weapons that few nations can match.

U.S. military superiority is most visible in air and naval power. The United States currently has over 6,000 military aircraft, nearly a third of the world inventory of combat aircraft.

U.S. aircraft are relatively long-legged, too; they generally have the longest combat ranges of any aircraft in the world. And there are some categories of U.S. aircraft that few other nations have. Other nations simply have no equivalent to our long-range bomber force, which today consists of 54 B-1B Lancers, 49 B-52H Stratofortresses, and 10 (out of a planned total of 21) B-2 stealth bombers. Nor can any single nation, or, for that matter, any combination of nations, match either the U.S. airborne tanker fleet of 54 KC-10 and 229 KC-135 tanker aircraft, which allow other aircraft to fly to longer ranges through airborne refueling. Only the U.S. military can sustain its aircraft globally. No other nation has our ability to operate aircraft virtually anywhere in the world because no one else has large aircraft carriers, a robust overseas system of air bases, and the support equipment to rapidly establish air bases where none exist.

Our tactical air arm remains second to none. These aircraft include Air Force airplanes such as the A-10 Warthog, designed as a close air-support tank killer, and the multi-mission F-15s and F-16s used for both air defense and ground attack. Similarly, the Navy's F-14s and F/A-18s that operate from aircraft carriers are designed for both offensive and defensive operations. This category also includes the Marine Corps's F/A-18D and the vertical takeoff and landing AV-8B Harrier. The list of tactical aircraft is not really complete without the Army's and Marine Corps's attack helicopter fleet, including the AH-1 Cobra and AH-64 Apache.

The United States has been a maritime nation throughout its history and a worldwide naval power since 1898, so it's not surprising that we have maintained a modern, high-technology naval force. The U.S. Navy's combat ships constitute about half of the total tonnage of all combat ships in the world. No other nation has the equivalent of a nuclear-powered, *Nimitz*-class aircraft carrier. There is no match for a U.S. surface ship equipped with the Aegis combat system,[28] or for the most modern U.S. Seawolf-class nuclear attack submarines.

Expense

Americans have spent a lot on their military—about $1.3 trillion over the last five years—and somewhere on the order of $4 trillion since 1979.[29] In recent years we have spent about $270 billion a year, which comes out to about 3.1 percent of the United States gross domestic product. We also allocate a relatively large portion of the U.S. national budget—about 16 percent—to the military. In Germany, Japan, and Great Britain the share of the national budget going to the military is less than 10 percent. The portion of the Iraqi national budget that goes to the Iraqi military is probably on the order of 30 percent or more. Russia and China probably spend about 12 percent. But no nation comes close to matching the $270 billion or so we spend annually on national defense. We spend about twice what all the nations of Europe spend together; about four times what Russia and the rest of the former republics of the Soviet Union spend; and about twice what the nations of the Middle East, Asia, and Southeast Asia all together spend on their militaries.

The defense budget in 1986 was about $282 billion. In constant 1998 dollars, however, it was about $400 billion, and this is the base figure the

Pentagon uses when it speaks of a budget decline on the order of about 40 percent since the mid-1980s.

We spend almost all the money that goes for defense on four things: paying the people in the military (personnel accounts), buying the equipment and weapons they will use (procurement accounts), operating and maintaining the forces and their equipment (operations and maintenance, O&M, accounts), and research and development (R&D accounts). About two-thirds of the budget goes for personnel and O&M; the remaining one-third is split more or less evenly between procurement and R&D. In general then, about two-thirds of the $270 billion or so spent on defense each year goes to run the existing force—to pay the men and women in the military and pay for what they do—while the remaining third goes into procuring the weapons and systems to build the future force. So to use the Pentagon budget as a guide to where the military is heading, it's good to focus on what is going into procurement and research and development.

The procurement budget buys major new big-ticket weapons systems, such as the B-2 stealth bomber, the Seawolf nuclear attack submarine, the F-22 Raptor third-generation stealth fighter, or the Army's planned RAH-66 Comanche scout helicopter. These highly visible aircraft and weapons generally get the most attention and scrutiny by the Congress, the news media, and the American public, and, unfortunately, these weapons systems are often given priority over new information technologies earmarked for military use. Much of the annual procurement budget goes to "recapitalize" the inventory of our stocks of existing equipment, including everything from trucks to tent pegs. The procurement budget required each year is therefore partly a function of the size of the force and its activity. Maintaining a large force at high levels of combat readiness—which means, among other things, conducting continuous, intense unit training—and operating the force at a high operational tempo conducting real-world deployments and missions—wears equipment out quickly. The inventory has to be replaced unless we are willing to either reduce the force size or its tempo of operations.

Over the last decade, and particularly over the past five years, spending on procurement has declined much faster than other portions of the defense budget. When the Defense Department began to shave down the size of the Army, Navy, and Air Force following Operation Desert Storm, the forces could live off of the budget largess of the 1980s. Given the political consensus that each of the services should be shrunk by an equal fraction, throughout the military the oldest equipment was retired first. So as late as 1993 we were still replacing old equipment with new equipment that had been budgeted for before the collapse of the Soviet Union in 1989.

But the "buffer" provided by the largess of the 1980s is now long gone. For the last three years the Pentagon procurement budget has run at least $20 billion less than what was necessary to replace worn-out equipment, from tent pegs to tanks, ships, and aircraft. We are beginning to see the effects of this budget trend. Combat units are less than ready, and, more important, far less well-equipped than they could be. And as a result morale is suffering. In short, we are once more facing the threat of what former Army Chief of Staff General Edward "Shy" Meyer warned against in the 1970s: "a hollow military" that on paper appears sound and effective but is actually short of people and modern equipment and lacks the ability to carry out combat missions effectively. If we don't change some part of this equation, once again we will have a relatively large military, composed of insufficiently trained personnel, who must struggle against outdated equipment to carry out their mission. It will be a military that will still be able to mount impressive parades. It just won't be able to fight very well.

There are short-term remedies, of course. We could increase the annual Pentagon budget by 10 percent. We could reduce the size of the military some more and cut back on its missions overseas. We could close more military bases. And we could reduce the profit margins of defense contractors.

But all of these steps are easier said than done. Without a clear threat, Congress and the American people won't support a massive hike in defense spending; but neither will they allow the United States to scale back our humanitarian missions and peacekeeping operations, no matter how much they cost.

It's true that the Pentagon has dealt with these very issues repeatedly since the end of World War II—balancing the demands of procurement, readiness, operations and maintenance, force structure, and end strength. And it has procedures in place designed to pound out the compromises needed to forge the annual budget requests. Something crucially important has changed, though.

The basic consensus on future defense needs that once knit together the various interests inside the Department of Defense is no longer there. And the concept of stewardship that permeated earlier generations—the commitment to preserve what their predecessors had passed to their charge—is eroding under the nagging suspicion that what those earlier generations designed and built for that earlier military era is wrong for the new one.

COMPARING AND ASSESSING FORCE STRENGTH

There are simple budget issues that define the trouble facing the U.S. military: not enough money; too many aging aircraft, ships, and weapons; and a grueling pace of current operations that is taking a toll on people and equipment. But identifying the hard choices ahead requires that we study one technical aspect of military strength that rarely appears in the budget debates.

People, equipment, and dollars, by themselves, reveal little about a nation's true military strength. Only by organizing these three components into comparable combat units can we start describing contemporary U.S. military power. The Defense Department has a lot to say about the categories and terms it uses to combine people and equipment into lists of conventional forces. And the specific manner in which it presents this information tells us a lot about the military institution's basic assumptions about what actually does constitute military power. This is a point that becomes very relevant in this book as we examine the technological elements of the emerging Revolution in Military Affairs.[30]

The Defense Department usually describes and measures the nation's land forces in terms of *organizations.* For example, a basic element of ground power used by the Army and Marine Corps is the division. And since that organization can vary a lot in terms of the particular mix of assets that make it up, comparisons of the number of brigades, battalions, companies, or other component units are really only an indirect measure of actual combat power in the division. So one has to be careful about counts and comparisons of military forces in terms of divisions, squadrons, or any of the other unit measurements one runs into so often.

U.S. Army and Marine Corps divisions are a case in point. The active component of the Army has ten divisions, and the active component of the Marine Corps has another three divisions. So the United States has, overall, a land force of thirteen divisions, right? Well, not really. What sounds like a pretty straightforward listing—the division—actually denotes two radically different organizations in the two ground services.

While the Army sees a division as its primary combat organization, the Marine Corps sees its "division" as an administrative organization. Thus you have Army divisions with a roster of between 15,000 and 18,000 personnel serving in component brigades of about 3,000 troops each, while the Marine division, with manpower totaling over 40,000, denotes a major concentration of personnel for management and administrative purposes. The Marine Corps's combat building block is what the force calls a Marine Air-Ground Task Force, a fourfold entity consisting of a command element, a ground combat component, an air combat component, and a combat service support element; sized for a particular mission, this task force can range in size from a Marine Expeditionary Unit of 2,000–3,000 Marines to a Marine Expeditionary Force of over 50,000 Marines consisting of a full division with air and logistical components added.

Nor are Army divisions themselves uniform in size, manning, and equipment. The Army in 1999 had six "heavy" and four "light" divisions in the active force. Light forces such as the 82nd Airborne Division and the 101st Air Assault Division are tailored for rapid, forcible-entry operations and for operations on restricted terrain and carry a relatively light load of equipment. Heavy units such as the 1st Cavalry Division or 4th Infantry Division are designed to fight other armies equipped with heavy weapons and are themselves armed with M1A2 Abrams tanks, M3 Bradley tracked infantry fighting vehicles, and M109 self-propelled artillery.

Likewise, the Pentagon budget and other documents assess naval power by focusing on types of warships: aircraft carriers, cruisers, destroyers, or attack submarines. The results are often misleading owing to the considerable difference in military capabilities among ships of the same type even in the same Navy. For instance, a modern *Arleigh*

Burke-class guided missile destroyer—armed with ninety vertical-launch cells that can carry both Tomahawk land-attack cruise missiles and standard antiaircraft missiles, and equipped with the Aegis anti-warfare radar and fire-control system, is significantly more powerful than the first-generation Spruance destroyers, which could carry only eight Tomahawks and did not have radar systems as powerful as the Aegis system. Yet both ships were lumped together in the "destroyers" category.

The roster of combat aircraft also presents an ambiguous picture of force size and strength. The U.S. Air Force last year reported that it had 113 heavy bombers available for combat (including 48 B-1B Lancers, 10 B-2 Spirit stealth bombers, and 44 B-52Hs). But the fleet actually consists of 73 B-1Bs, 21 B-2s, and 85 B-52Hs. The difference is one of budgetary bookkeeping: The larger numbers represent the "total active inventory" of operational aircraft including those assigned to the reserves, training, long-term modifications, or test and evaluation programs, while the smaller number in the "primary aircraft inventory" category constitutes those aircraft directly assigned to active combat forces.

The Department of Defense in one recent tally said there were 2,280 fighter/attack aircraft in the active force.[31] Throughout the four services, there were at last count about 2,300 *additional* fighter/attack aircraft that were being maintained for training, testing, attrition replacements, and reconstitution reserves. And roughly 1,000 additional aircraft have been mothballed (put into protected storage) and could be brought back into service in a matter of weeks to months.

As these examples show, the inaccurate criteria long employed by defense planners for gauging military strength is a vital clue to the obsolescence of the current U.S. military force structure. The charts and columns of figures reflect a decades-long tradition of assessing combat power by different types of *platforms*—aircraft type, warship classes, and specific pieces of Army or Marine Corps ground equipment—while ignoring the more precise (even if difficult to tabulate) compilation of combat power based on the ability to place weapons on target.

The differences that separate the Army, Navy, Air Force, and Marine Corps from one another are not intrinsically bad. In fact, they reflect the historically different missions assigned to each service, and the impact of the physical environment on how the organization has been structured over decades of experience.

The Army and Marine Corps are both designed to seize and hold territory, and to destroy enemy formations challenging them for that terrain. Even with that common mission, it is obvious that the two ground forces will be designed differently and will train and fight under different doctrines. Historically the Marine Corps—originating as a naval infantry auxiliary of the Navy—has focused on amphibious attacks from the sea such as its series of assaults in the Pacific during World War II. It has structured its forces to operate on naval amphibious ships, and to go ashore with a self-sustaining air arm as well as its own supply organization.

Although the Army, too, has a "forcible entry" capability with its elite Ranger battalions and "light" divisions, the Army in recent decades has adapted to a strategy using U.S. Air Force transport aircraft to carry initial waves of paratroopers and supplies, and a combination of sealift and aircraft to ferry the "heavy" divisions to distant war zones.[32]

The Navy and Air Force retain the specific mission areas for which they have been historically known. The sea service operates a "blue water" fleet capable of carrying out combat operations against other naval powers in the deep ocean while expanding its doctrine to embrace combat operations in the littoral regions (including a new emphasis on waging war ashore using aircraft and missiles), supporting the Marines with its amphibious fleet, and carrying out its share of the strategic nuclear deterrent mission with its fleet of 18 Trident missile submarines. The Air Force remains the primary strategic air power with long-range bombers and a large tactical fighter inventory. It is the prime airlift force with its fleet of long-range cargo aircraft and aerial refueling tankers.

Yet it would not be difficult—and perhaps it would be more informative—for the Pentagon to devise a set of common criteria for describing combat power residing in the Army, Navy, Air Force, and Marine Corps. All four services do a lot of the same things, and a lot of them similarly. Because they all move, shoot, and communicate, we should describe and compare the forces in terms of how and how well they can carry out those essential functions. Similarly, because they all train, feed, pay, clothe, and promote the people who wear their distinctive uniforms and insignia, we could portray and measure them in terms of how they do those things. And since they all acquire and maintain equipment (much of it essentially the same equipment, provided by the same producers), the use of equipment could form a common measurement. Finally, since they all fight wars, provide military presence, and train allies, there are enough similarities across the board that it seems easy to draft some common metrics with which to describe and compare how the different services operate. But we don't—not today.

Why not? I think the different measures we use in talking about our force components are much more a product of tradition and interservice rivalry than they are a result of logic or of a sense that these components together contribute to a joint military capability. We use different measures—without much thinking about it—because the military services have wanted to maintain and emphasize their differences, and because, over the years, Congress and the electorate have accepted the notion that each of the military services is a separate, discrete, and fundamentally different institution.

CORPS IDENTITY

There is some value to highlighting the differences and distinctive aspects of the different forces. A sense of distinctiveness encourages *esprit de corps*, bolsters discipline, and fosters valor in battle. It helps to weld individuals into a team, to increase particular skills and specialties, and to enhance unity. Such strengths are very important to military forces and to their combat effectiveness.

In the Navy, many young sailors who opt to reenlist for another four-year tour often arrange the brief ceremony at a location evocative of the long history of sea service. One of the more popular sites is the "surrender deck" on the retired battleship USS *Missouri*, the exact location where Admiral Chester Nimitz and General Douglas MacArthur accepted the Japanese surrender to end World War II on September 2, 1945. Army soldiers are reminded of their rich history every day simply by regarding their unit patches (fifty-six years after Operation Overlord on June 6, 1944, most of the units that stormed the Normandy beaches remain on active service, from the U.S. Army V Corps in Germany, to the 1st, 4th, 82nd, and 101st Divisions, to the 2nd Ranger Battalion).

But there is a darker side to this celebration of unique military identity. It cannot help but encourage a separation of Army from Marines, of Air Force from Navy. The historic and (military) cultural traditions that foster this separation promote an excessive and expensive redundancy, including separate medical, intelligence, and logistics organizations. And despite a decade and a half of laws and policies encouraging multiservice cooperation and joint operations, the traditions of the forces can erode the effectiveness of joint operations and impede the synergy of true military cooperation that is essential to carrying out modern combat operations. Some examples of this tendency in the recent past include the Army's and Navy's inability to communicate with one another during the 1983 Grenada invasion, and the incompatibility of Navy–Air Force communications systems during Operation Desert Storm, which severely complicated joint air missions. (The role interservice rivalries and misunderstandings have played in outright U.S. military failures is examined more fully in Chapter 2.)

THE CHALLENGES AHEAD

The U.S. military today is poised halfway between the industrial age and the information age that has succeeded it. At century's end the armed forces continue to be designed to fight war as it has evolved during the industrial age—an age that has already ended. Powerful new information technology applications that are transforming civilian society and are ushering in a global economy are also entering the military. The open issue before us is whether our civilian and military leaders can apply the new technology to transform the U.S. military into a twenty-first-century fighting force that is inferior to none, or whether the new technology will be misapplied, leaving the force smaller and less capable than before.

During the Cold War, the United States had the luxury of a general consensus on what the most important national security problems were and what to do about them. There was considerable debate over the details, and professional careers sometimes rose and fell over how those details were decided and implemented. But virtually everyone in national security decision-making positions, from the early 1960s to the early 1990s, recognized the same threat and shared most of the same assumptions about what was to be done. So we built a major conventional and nuclear force and aimed it at a superpower rival.

We built and shaped, revised and honed, trained and prepared our military forces within this particular template. We based our strategy and operations on our sense of Soviet military capabilities and postulated global war centered in Europe. Year after year, inside the Pentagon and its supporting think tanks, in the fleets and in the field, in the military schools and war colleges, we worked out what such a conflict meant for the size, structure, and character of U.S. military forces. But we became trapped inside our own assumptions. So much so that when the Soviet Union began declining in the late 1980s and collapsed altogether in 1991, most of the U.S. defense establishment was caught by surprise. That lethargy explains the caution with which we entered the post–Cold War era.

The challenge before us today is heightened by the limited amount of time experts estimate we have to carry out the new American Revolution in Military Affairs: the "strategic pause" that began with the collapse of the Soviet Union in 1991, defined as a grace period during which American military power will remain unchallenged by any "peer" competitor, is expected to last only until 2010 or 2015 at the most.[33]

Our predicament is all the more urgent because of the long gestation period that is necessary for developing and testing modern weapons and combat systems. The aircraft carriers, tanks, and aircraft in the force today were designed more than twenty years ago. The F-16s that are the most numerous tactical aircraft in the U.S. Air Force today were designed in the early 1970s and procured in the late 1980s. The Army's M1A1 main battle tank was designed in the late 1960s; procurement began in the 1970s.

When the aircraft carrier USS *Abraham Lincoln* reached its new home port in Everett, Washington, in 1997, it celebrated a double anniversary. It had been exactly fifteen years since the carrier keel had been laid, beginning construction; fifteen years since the new base had been announced in 1982.

Contrast that glacially slow chronology with the explosion of computer technology and the rise of the Internet in the past decade. Imagine doing your office work with a computer designed twenty or thirty years ago. Imagine running your business with a fleet of 1970s-era Ford Fairlanes. Try imagining the current U.S. military structure coping with Moore's Law—the unofficial theorem that off-the-shelf computing power doubles every eighteen months.[34] Those are the challenges facing us as we confront the technical and political obstacles that must be overcome to successfully implement the Revolution in Military Affairs.

CHAPTER 2

SEEDS OF REVOLUTION

I was staring out the cockpit window of the Navy jet at the massive warship steaming below when I suddenly realized that the U.S. military had passed from the Cold War into an entirely new era.

It was the summer of 1991, several months after a U.S.-led military coalition had driven the Iraqi Army from Kuwait, and the attention of U.S. military leaders was once more focused on the Soviet Union and its ongoing political and military turmoil.

I had assumed command of the 6th Fleet in the Mediterranean Sea in August of 1990 at a time when we were getting contradictory and mixed signals from our longtime adversary. While signs of profound change within the Soviet Union gave hope that the Cold War might be coming to an end soon, in mid-1991 that former superpower's armed forces were still the single greatest potential threat to the United States and NATO. Thus, on orders from the Pentagon and NATO headquarters, my command continued to define our top priority as deterring Soviet aggression. The 6th Fleet was still a powerful force in the early 1990s—one of four numbered fleets covering the world's trouble spots. Routinely there were several U.S. aircraft carrier battle groups and amphibious task forces carrying Marines and nuclear attack submarines on peacetime patrols in the Mediterranean. The ships, aircraft, and personnel under my command were still organized, tasked, and trained to be ready for war against the Soviets, whatever the headlines might say.

But change was undeniably in the wind. My arrival at fleet headquarters in Gaeta, Italy, in August 1990 came just a year after historic political changes had swept through Eastern Europe as the once-solid Warsaw Pact began to disintegrate and its member countries grappled with an outpouring of popular demand for freedom. And from the outset of the massive U.S.-led military buildup in the Persian Gulf that began during my first months as 6th Fleet commander, the government of Mikhail Gorbachev signaled a historic reversal when it abandoned Iraq, its longtime Middle East ally, to offer political (but not military) support for the allies in the Persian Gulf War.

And now hard evidence of the new strategic era lay before my very eyes.

Several days earlier an electrifying message had arrived at my headquarters in Gaeta. After more than a decade of construction, fitting out, and sea trials, the Admiral Kuznetsov, *the Soviet Union's first full-sized aircraft carrier, had finally left port at Nikolayev in the Black Sea and was heading into the Mediterranean.*[1] *I had flown out to the aircraft carrier USS* Saratoga, *then operating in the eastern Mediterranean, and a half hour earlier had climbed into an S-3A Viking antisubmarine patrol jet and gone up to 6,000 feet to get an overhead look at the Soviet carrier. What I saw seemed at first glance to confirm the Pentagon's long-standing fear that the Soviet Navy was ready to challenge our navy for supremacy at sea. Yet after a closer look I realized that not only was the Soviet warship no great achievement, but that the decades-long military rivalry between the two superpowers had entered a terminal phase.*

For decades the Soviet Navy had put most of its naval effort into building a mammoth fleet of nearly 350 conventional and nuclear attack and missile submarines, prompting the United States to greatly increase its nuclear submarine construction in response. Now, although the Pentagon judged that the Soviet Navy's mission was meant to defend the USSR's coastlines from attack, the presence of the Admiral Kuznetsov *suggested that despite its economic woes the Soviet Navy was still determined to become a "blue water" force that could challenge the U.S. Navy worldwide.*[2]

But on this morning in the summer of 1991, as I stared through binoculars scanning the steel flight deck of the Admiral Kuznetsov, *I saw something quite different. The carrier was not a combat-ready warship at all. Except for a pair of unarmed helicopters there were no combat aircraft in sight. The flight deck and hull were streaked with rust. Several women were walking their pet dogs on the flight deck.*

When it steamed out toward the Mediterranean, the Soviet carrier was not preparing to challenge anyone. It had literally been evicted from the Black Sea upon the secession from the Soviet Union by Ukraine, which as a newly independent nation had taken possession of most of the harbors and naval facilities in the Black Sea that had previously been the home port of a major Soviet fleet. Rather than deploying to confront the U.S. Navy, the Kuznetsov *was on its way to its new home port far above the Arctic Circle. I ordered a number of the 6th Fleet's destroyers and cruisers to form a friendly escort for the* Admiral Kuznetsov *on its trip across the Mediterranean, and radioed its commander, Captain Sergei Chekhov, to wish him the traditional greeting of "fair winds and following seas" on his voyage home.*

My former adversary radioed back, "I no longer know where my home is."

What made that moment so significant? It wasn't simply a sign of the end of the Cold War: by 1991 those signs were everywhere. No, my bird's-eye view of the rusty Soviet aircraft carrier on its way to the far, far north struck me as significant because it spelled an end to a long military rivalry—indeed, the end to a way of thinking about war that has been in place longer than any of us can remember, that for decades shaped our politics, our economy, and our sense of the proper role of the United States in the world.

The 1990s was a decade of multiple revolutions—political, economic, technological—that changed so thoroughly the way we live that the past no longer seems a good guide to the future (in fact, the past seems precisely the wrong guide). So it is in the world of military affairs. The Revolution in Military Affairs is our opportunity to use the new information technology to change the very nature of our military, in a way that could reinvigorate American political, diplomatic, and economic leadership in the world for decades to come. But for this revolution to take place, the people in charge of the military—and the people they serve, namely, the citizens, the Congress, and the President—will have to begin to think in a radically different way about war and peace, about armies and navies—and they have to start thinking differently right away. There have been many more examples like the voyage of the *Admiral Kuznetsov* since 1991 to remind us that the changing world demands a new way of looking at war and the proper military force for the United States in the new century.

As noted in Chapter 1, this revolution hangs on the U.S. capability to harness new generations of information technology to transform the very nature of military power. For example, it means enhancing our current early-warning systems—currently based on radar—with an elaborate network of satellites that can employ a wide range of sensors to give our forces better information and protect the United States and its military units from attack. But the Revolution in Military Affairs is about more than new weapons and technologies. It will require that we be ruthless in reexamining the root premises that define the U.S. military today: the overall size of the force, the current roles and missions of the four combat services, the doctrine that guides how we decide on which weapons to build. It will force us to look again at the way the Pentagon, Congress, and the White House determine the massive annual defense budget. And it will force us to reexamine from the top down the fundamental strategy that governs the way in which our nation actually wages war.

Can we do it? It depends on our ability to recognize the imperative for significant change, and our willingness to abandon obsolete ways of thinking about national defense. Because for the revolution to really be a revolution, the military will have to be significantly reorganized. The way in which the Pentagon reorganizes the military to best use the technology will be as critical to the success of the Revolution in Military Affairs as the new technology itself.

Ranging from space-based satellites and unmanned air and ground sensors to secure global communications and an arsenal of precision-guided weapons, the elements of the military revolution constitute the specific means of providing the United States with an unsurpassed military. Less visible but equally important are the "enabling" components of the revolution: the shifts in military organization, command structure, communications, and training that will connect the technology to the people who will use it in combat.

One point is very clear: the Revolution in Military Affairs is happening whether U.S. political and military leaders endorse it or not. From China to Western Europe, nations we count as allies and adversaries alike are aware of the promises of military transformation using information-age technology linked to modern weapons. Weaker nations are studying how to use technology to neutralize or offset the overwhelming military power of the United States. The issue is whether we as a nation have the imagination and courage to take the steps necessary to preserve our current superiority.

THE SHAPE OF MODERN WAR

To understand both the promise of the new Revolution in Military Affairs and the deep challenge we face as we seek to carry it out, it is instructive to look at the nature of war in the modern age, as well as how earlier military revolutions in history have transformed a nation's ability to survive and prevail against its adversaries. As we will see, the current Revolution in Military Affairs is not the first that has swept the U.S. military in our nation's history.

Until recently most people in the developed world understood war in the same way: as an organized military conflict between two hostile nation-states, sparked by competing claims for territory or markets, religious or ideological hostility, or a perceived need for self-defense against the threat of others—the long-standing troika of "gold, gospel, and glory." Today one may add to that some additional forms of warfare. Lately, we have seen wars of counterterrorism (as against accused Saudi terrorist Osama Bin Laden) and what are called "low-intensity conflicts" such as the clan warfare in Somalia that prompted the 1992 United Nations intervention in the Horn of Africa. Both types of conflict involve the use of military power against subnational groups (terrorists, organized criminal cartels) that in one way or another are threatening the lives of American citizens or U.S. national interests. In any event, war today still involves a nation's use of organized violence against an adversary to attain the goals set by the national leadership.

In this century military planners (such as the men I studied under at the U.S. Naval Academy) have come up with four different levels of war.

The first is *total war.* Total war is a conflict in which all of the resources of the nation and its citizens are mobilized to defend against a threat perceived to jeopardize the very existence of the home country. The most obvious examples of total war in our time are the U.S., British, and Soviet mobilizations to fight Nazi Germany and imperial Japan during World War II. After Hiroshima and Nagasaki, theorists came up with the term *strategic nuclear exchange* to describe an apocalypse in which the United States and Soviet Union would launch their long-range nuclear weapons at one another so to destroy one another—fortunately that kind of "total war" didn't come to pass.

The second level of war is known as an *unrestricted conventional war.* This kind of conflict involves a massive use of conventional arms—air strikes, ground forces, and naval attacks against another nation—but not nuclear, chemical, or biological weapons. Generally such a war—fought with or without the support of military allies or coalition members—would call on the military to confront an adversary that threatens vital national interests but not the life of the nation itself. This kind of war is also called a *major theater war* or *a major regional conflict.* The most recent example of such a conflict is the 1991 Persian Gulf War, in which the United States deployed about half its military force.

The third level of warfare is known as *limited war* or as a *contingency operation.* A substantial military force—usually no more than one or two Army divisions, Air Force wings, or a major part of a naval fleet—is brought to bear on a local crisis area to safeguard national interests and save civilian lives under the leadership of one of the multiservice U.S. military commands. The 1983 U.S. intervention in Grenada and the 1989 American invasion of Panama are good examples.

The fourth level of military engagement is known as a *low-intensity conflict.* Such operations take place in a vast gray area between peace and overt warfare. In the current era a low-intensity conflict can embrace military assistance to insurgent groups fighting against a nation the United States opposes, such as the covert support we provided to the Afghan rebels to fight the Soviet military occupation of Afghanistan in 1979–88; it can also involve support to embattled friendly governments threatened by insurgents who are unfriendly to the United States, such as our current low-level military support to Latin American countries such as Venezuela and Peru against leftist rebels allied with drug cartels. A variation of the low-intensity conflict is "gunboat diplomacy"—an extended show of military support short of direct combat participation for one side engaged in a conflict—such as the U.S. decision in 1988 to temporarily "reflag" Kuwaiti oil tankers as American vessels and to provide U.S. Navy escorts for the tankers in the Persian Gulf to deter Iranian attacks on the ships. Yet another common form of low-intensity conflict involves the quiet and sometimes secret use of U.S. Special Operations Command units in a variety of training assistance missions to friendly countries such as Indonesia in the early 1990s, or participation in direct combat strikes such as counterterrorist missions against Libya and Palestinian terrorists in the 1980s and 1990s.

A fifth category of military involvement—*operations other than war*—has emerged in the post–Cold War era. This category runs the gamut from peacekeeping missions such as the one undertaken by the multinational force sent to Bosnia to enforce the Dayton Treaty in 1995; to the U.N.-sponsored intervention in Somalia in 1992 (which began as a peacekeeping and humanitarian operation but turned into direct U.S. combat against one of the larger ethnic clans in Somalia); to major humanitarian rescue operations such as the U.S. Marine Corps response to a typhoon in Bangladesh in 1992 and the massive U.S. military aid mission to Central America in 1998 after Hurricane Mitch.[3]

Since it emerged as a world power a century ago, the United States has waged war in all of these ways. In general, our military objectives have been the same from one war to the next (we hoped to defeat the enemy and to deter future conflict), and our means changed to suit changing national interests or to match different enemies.

But on a number of occasions the U.S. military has changed its way of waging war in order to take advantage of great technological innovations, which turned out to change the nature of war itself. The submarine enabled us to destroy Japan's ability to attack us during World War II, when we strangled Japan's economy by destroying the nation's navy and merchant fleet (its economic lifeline); the atomic bomb provided a final series of shocks to persuade the Japanese high command to surrender. A large nuclear arsenal enabled us to "defeat" the Soviet Union in the Cold War by neutralizing its overwhelming superiority in conventional weapons in Europe. Our superiority in air power, together with a combination of sensors and electronic warfare capabilities, led us to blind the Iraqi Army to our movements and set the stage for defeat on the battlefield in Operation Desert Storm.

In these experiences, one important reality has emerged over and over again: New technology alone cannot bring about a revolution in military affairs. *But technology, if used properly in conjunction with the correct form of military organization, can strengthen a nation's military position, transform its ability to safeguard the national interests, and prevent the military from being ruined economically by the diversion of economic assets to wasteful or inefficient military programs.* In short, military history can instruct us as we contemplate how best to bring about a new Revolution in Military Affairs.

EARLIER MILITARY REVOLUTIONS

One challenge for U.S. military leaders and defense planners is to recognize the difference between ongoing change and a full-throttle Revolution in Military Affairs. Even in an era of tight budgets, the U.S. military today is continuously modernizing its combat systems and adding individual new weapons and systems as they emerge from the assembly line. For example, beginning with the USS *Virginia*, a new class of Navy nuclear attack submarine is being constructed whose design incorporates the lessons learned from previous models such as the Los Angeles and Seawolf classes. The Army is seeking to build a new Crusader self-propelled howitzer to replace earlier M109 Paladin weapons. The Air Force is well on its way to replacing its 1960-era fleet of C-141B transports with the C-17 Globemaster III. Each of these weapons systems constitutes a significant gain in performance or firepower, and each new weapon or system incorporates many elements of the computer revolution, but such advances in my judgment still fall short of the true potential of the Revolution in Military Affairs.

As I see it, the revolution means creating a *synergy* in new weapons, sensors, and communications that is made possible by the successful melding of the technological applications with an information-age military organization.

"In evolutionary change, progress is made by improving the last generation of military equipment and organizations, but continuity still exists between the old and new generations," military analyst Colonel Douglas Macgregor recently wrote. "In periods of revolutionary change, almost no continuity exists between generations—we are looking at something entirely new."[4]

The respected military historian Martin van Creveld divided military history into four ages: the "Age of Tools," the "Age of the Machine," the "Age of Systems," and the "Age of Automation."[5] In each of these ages, the nature of military power was suddenly transformed by the advent of new technology—and by the deployment of flexible, innovative military forces, which swiftly reconceived themselves so as to use the new technology in battle.

During the "Age of Tools," which ran from prehistoric times until about 1500 A.D., military force depended on energy from the muscular power of men and animals, augmented by what we can call the "leading edge" technology of the time of primitive technological developments—the wheel, the stirrup, and iron weaponry, which were put to use in battle in horse cavalry, chariots, and the sword.

In the twelfth century in Asia, the Mongols established an empire by taking advantage of a preindustrial Revolution in Military Affairs, applying the horse and the bow and arrow in new ways. In the sixty-eight years between the time Genghis Khan united the tribes of Mongolia in 1190 and the capture of Baghdad in 1258 by his grandson, Kublai Khan, Mongol horseback armies conquered northern China, Tibet, Korea, Central Asia, Persia, the Caucasus, Turkish Anatolia, and northern India. They struck in Europe from Russia to Austria.[6]

The Mongols were neither the first people to master the use of horses as military transportation nor the first to wage war with the bow and arrow—both "technologies" were by then hundreds of years old. But Genghis Khan was really the first commander to devise a way to use them, and to establish an entire military culture around them, with what one historian describes as "an efficiency and discipline never attained before."[7] In an era of clan societies, his innovations were revolutionary. Indeed, he appointed his army commanders based on performance and talent rather than hereditary position. He divided his armies into standard formations of ten, one hundred, one thousand, and ten thousand warriors. He used scouts to survey the enemy force's location, movements, and points of vulnerability.

But it was in Mongol battlefield strategy and tactics that Genghis Khan and his descendants were true military revolutionaries. Historian Bevin Alexander attributes Mongol dominance in part to their decision—stemming from their ethnic and historical experiences as nomads from the great Asian steppe—to limit their force to one major branch, the horse cavalry. By playing to his culture's strength and choosing not to field a large infantry army as well, Genghis Khan created a military force whose impact was felt from the Sea of Japan to the gates of Vienna. As Alexander recounts:

> The favorite Mongol tactic . . . was to use the "mangudai," a specially selected unit that would charge the enemy alone. After a fearsome assault, the mangudai would break ranks and flee in hopes of provoking the enemy to give chase. It was usually so convincing that the enemy cavalry sprang after the fleeing Mongols believing they were on the verge of victory. Unseen to the rear, Mongol archers waited. By the time they were upon the archers, the enemy horsemen were spread out and many would fall to well-aimed bow shots. Now disordered and suffering heavy casualties, the enemy would be vulnerable to the (Mongol) heavy cavalry, which now made its charge.[8]

According to defense analyst James Adams, revolutionary developments in this early age of warfare included the widespread appearance of the bow (twelfth century); the transformation of siege warfare against fortresses and walled cities with the appearance of early artillery weapons (seventeenth century); the first placement of artillery on ships (also seventeenth century); and the proliferation of infantry armed with muskets.[9]

During the second age, the "Age of the Machine," Western civil societies changed from agricultural to machine economies, and their armies and navies changed accordingly—as was seen in the Napoleonic Wars and our own Civil War.

Napoleon, a French general who rose to ultimate political power in 1799, seven years after a Prussian-Russian alliance had invaded his country, is usually given credit for tapping into the French Revolution to transform the relationship between the citizen and the military. But he exploited new technology as well. Just four years into his leadership, Napoleon was able to raise a mass army of more than a million men, dispatching them on conquest from Spain to Russia. Military historian John Keegan writes that Napoleon created a citizen army based on the revolutionary concept that "accepted the right of the state to demand the duty of every fit male individual to render military service."[10] But he was able to do so because of an economic transformation in French society that made possible the mass production of weapons in factories. "Napoleon brought a revolution in military affairs by assimilating the weapons technology of the age into a consistent pattern of military theory, organization and leadership," Macgregor writes. "This congruence of French weapons, tactics, organization and thinking about war reflected Napoleon's understanding of how to use existing technology to the limit and at the same time make its very limitations work to French advantage."[11]

In order to organize such a mass of armed men, Napoleon conceived the first modern field army organization, the *army corps*, with a commander leading between two and five divisions with up to 100,000 soldiers in all. Nearly two centuries later, the corps remains the primary organizational unit of ground war in all modern armies, including ours.[12] Napoleon capitalized on recent advances in artillery weapons design by forming independent, mobile cannon units to support the larger infantry formations, his tactics heralding the shift in emphasis from *manpower* to *firepower* on the battlefield. Napoleon's transformation of the 200,000-man French Army into a force of six corps of between 25,000 and 30,000 soldiers apiece gave him six "mini-armies," more than quadrupling the number of combat units available in a similar-sized force at that time.

Napoleon's revolution in military technology did not make war any less of a horror for those involved than it was for Genghis Khan and his contemporaries. In fact, the industrial age made war a mass-production slaughterhouse with casualties escalating by a factor of ten for both military participants and civilian bystanders. The full horror of mechanized combat would become apparent in 1861, thirty-five years after Napoleon's defeat at Waterloo, when the U.S. Civil War broke out with the Confederate bombardment of the U.S. Army at Fort Sumter in Charleston, South Carolina. Many believe that industrialization played a defining role in the Civil War. Given its industrial base and its 2–1 advantage in terms of population, the Union started the Civil War a heavy favorite. The Union army—transformed in fifty years from a rural constabulary scattered in isolated garrisons and territorial posts after the War of 1812 to an American clone of Napoleon's Grande

Armée fifty years later—owed its transformation to military applications of new technological breakthroughs then transforming American civil society. The Union made use of the newly developed wire telegraph and civilian railroads to control military operations and rush massive combat units into battle. Both sides hastened to employ the ironclad warship, first seen in the historic battle between the Union *Monitor* and the Confederate *Virginia* in Hampton Roads, Virginia, on March 9, 1862. And the South launched a true submarine, the CSS *H. L. Hunley*, to torpedo and sink the Union steam frigate USS *Housatonic* on February 17, 1865 (the mission proved fatal to the *Hunley* as well, which sank with all hands after its weapon—a contact explosive—detonated against the Union frigate).

Moreover, General Ulysses S. Grant recognized the need to revise military strategy in order to capitalize on his army's new technological superiority. Upon taking command of Union forces in 1864, Grant immediately reorganized his troops into a force Napoleon would have easily recognized: five separate commands under subordinate generals who were capable of waging independent operations but able to merge into an overwhelming single force if necessary. An admiring General William Tecumseh Sherman would write Grant, "That we are now all to act in a Common Plan, converging on a Common Center, looks like Enlightened War."[13]

The dark legacy of the industrial era came to light in the U.S. Civil War, during which armies used mass-produced weapons to fight a war of attrition and victory was defined as the destruction of the other side's fighting force and civilian economy. This approach to war led to casualties on a scale unheard of in recent history. In four years the Union and Confederate armies fought 2,261 battles with horrific casualties—395,528 Union soldiers out of about 2,000,000 participants were killed, a fatality rate of 18 percent; and there were about 258,000 Southern deaths out of a force of 1,000,000, for a fatality rate of nearly 26 percent—making it the bloodiest war in our history.

Subsequent major conflicts, particularly the two world wars of the twentieth century, would see even further escalations in the scale of violence perpetrated against civilian populations—in many cases as a matter of cold-blooded policy.[14]

The third military age, termed by van Creveld the "Age of Systems," occurred in the early twentieth century in the interval between the two world wars. Any history of World War II must focus on the military technology and weaponry that scarcely existed during World War I. Some examples are instructive.

Modern Aircraft

At the time of the U.S. entry into World War I—just fourteen years after the Wright Brothers' first flight at Kitty Hawk in 1903—American air power consisted of the Aviation Section of the U.S. Army Signal Corps, with fewer than 500 biplanes commanded by a lieutenant colonel and staffed by 1,218 personnel. These primitive aircraft were useful for daylight reconnaissance and daytime air-to-air combat against enemy aircraft and dirigibles—the famous duels of Baron von Richthofen and his like—but little else.

During World War II the Army Air Corps expanded from 51,000 personnel in 1940 to a worldwide force of 2.2 million men and women in 1945. It was new technology that made possible the development of long-range combat aircraft featuring more powerful engines and pressurized cockpits and fuselages that enabled planes to fly at altitudes in excess of 11,000 feet.

Radar

Few technologies have transformed the use of military force more than "radio detection and ranging" technology, which can pinpoint the position of distant aircraft and ships by transmitting a radio wave and recording its reflection off the target. Radar was first identified in the 1920s by American, British, and German scientists; by 1940 it played a key role in the outcome of the Battle of Britain: when German fighters and bombers based in occupied France and Scandinavia attempted to destroy the British will to fight by attacking both military and civilian targets, the British were able to use radar stations to detect and warn the public about the coming air raids. Although the Germans had three times as many planes as the British—and at times 1,000 German fighters and bombers operated over England in a single day—the Royal Air Force prevailed, broke Germany's resolve, and prompted Adolf Hitler to cancel his planned invasion of England.

Blitzkrieg

The most famous (and notorious) example of new technology wedded to innovative military organization and tactics was Nazi Germany's use of the blitzkrieg,[15] or "lightning war," which combined the use of close air support, radio communications, and German tank divisions specifically designed to outrace and outgun the enemy.

Before declaring war, German Army generals, particularly Hans Guderian, proposed coordinating a force of armored panzer units with a fleet of aircraft overhead, linked by a network of tactical radio systems. The aircraft pilots would spot the targets, then use the new FM radio to contact the tank commanders, who would then race their vehicles to the identified site and strike swiftly and effectively before the defenders could regroup. Precursor operations during the Spanish Civil War in the late 1930s, Germany's invasion of Poland in 1939, and Germany's invasions of Denmark and Norway in April 1940 validated the concept, and in May 1940 a six-week German blitzkrieg led the British Expeditionary Force to flee from France at Dunkirk in humiliation, and France to surrender. Used together, three new kinds of technology—fast tanks, high-flying planes, and reliable radios—formed a successful military strategy. Here is a key lesson about technology and war: synergy wins. Western opponents were unprepared for the consequences of a German revolution in military affairs.

The fourth age identified by van Creveld—the "Age of Automation"—is the age in which we now live: the Cold War and the post–Cold War era of ethnic strife and subnational terrorism.

ROOTS OF THE CURRENT REVOLUTION

The current American Revolution in Military Affairs did not appear hand in hand with a new weapons system or some other kind of military technology. The revolution's high-tech weapons—such as air-dropped precision-guided munitions and stealth aircraft—were first developed in the 1970s and have been modified steadily ever since. But the revolution has been spurred by the development of communications technologies that were not even imagined twenty years ago, such as computer processors and handheld Global Positioning System (GPS) satellite navigational receivers. And the future of the revolution will be defined by technologies we have only just begun to develop, such as hypersonic cruise missiles and electromagnetic pulse weapons.

Still, if one has to affix a date to the beginning of the present revolution, it is 1977, when three key Pentagon officials—Harold Brown, Andrew Marshall, and William J. Perry—began to think in concert about the application of technology to military affairs. Secretary of Defense James Schlesinger had named Marshall the Defense Department's first director of net assessment in the early 1970s. Coming into the Pentagon with impressive Soviet-watcher credentials, Marshall focused the analytic work of his office on the military relationship between the United States and the USSR—until the collapse of the Soviet Union. His work involved efforts to measure the military balance between the superpowers and shift it in favor of the United States. Mostly, this meant reading a lot about what the Soviets were saying to themselves.

After Jimmy Carter was elected President in 1976, Secretary of Defense Harold Brown brought Perry into the Pentagon as undersecretary of defense for research and engineering, a low-visibility post yet one that gave Perry a pivotal role in the U.S. military's astonishing and inspiring self-rejuvenation in the decade after the fall of South Vietnam in 1975.[16]

Secretary Brown's main achievement in his four years as Pentagon chief was to devise a focused program by which the United States and NATO allies could use Western technological superiority to neutralize the overwhelming size advantage that the Soviet Union and its fellow Warsaw Pact members had over NATO forces in Europe. The desire to move away from industrial war butchery was not new. During the 1950–53 Korean War, Army General James A. Van Fleet had noted the preference "to expend fire and steel, not men," when confronting the enemy, but, overall, weapons and systems of that era did not remove the probability that combatants would suffer attrition-war casualties.

Two decades later the "offset strategy" devised by Brown and his subordinates offered a real chance for technological quality to replace mass-on-mass quantity in battle.

But for Perry (as he explained in a 1991 article) the offset strategy was not merely a plan to ramrod high technology for its own sake into the U.S. armed forces:

> Much of this debate surrounding the offset strategy this past decade has been based on the false assumption that its primary objective was to use "high technology" to build better weapon systems than those of the Soviet Union. But it was not likely that higher quality tanks, guns, or aircraft, by themselves, could offset a three-to-one (Soviet) advantage. The offset strategy was based instead on the premise that it was necessary to give these weapons a significant competitive advantage over their opposing counterparts by supporting them on the battlefield with newly developed equipment that multiplied their combat effectiveness. This strategy was pursued consistently by five administrations during the 1970s and 1980s, all the while being carefully observed by the Soviet Union.[17]

Two key elements of the offset strategy were classified as state secrets and so received little or no publicity at the time. The first was an intense campaign to design and build "stealth" aircraft whose aerodynamic shape and construction materials were designed to make them invisible (or very nearly so) to Soviet radar systems. Design and engineering work on both the F-117A stealth fighter and the B-2A stealth bomber were begun during the late 1970s. The second element of the strategy was a program called "assault breaker," an effort to defeat Soviet armored forces with a variety of command, control, and intelligence systems, advanced communications, and precision-guided weapons.

In retrospect, the offset strategy must be recognized as an important precursor to the American Revolution in Military Affairs because of a key attribute: Rather than focusing exclusively on specific weapons systems, combat functions (such as long-range bombing or submarine operations), or service missions, Secretary Brown and Undersecretary Perry led a campaign within the U.S. military to strengthen the force through technological *synergy*. It did not take long for our Soviet adversaries to notice what the Pentagon was doing.

Marshall and his colleagues were intrigued in the late 1970s when Soviet military theorists started writing in their professional journals about a "military-technical revolution." Soviet technocrats began to argue that computers, space surveillance, and long-range missiles were merging into a new level of military technology, significant enough to shift the balance of power in Europe in favor of the United States and NATO. Many Soviet watchers in the West at the time thought such writings might be signals of new advances in Soviet military technology. But to Marshall, something else was at play. He reasoned that the Soviets were actually motivated more by anxiety over what the United States was doing than by technical smugness over Soviet programs. The Soviets, he suggested, were actually reacting to U.S. programs such as the "assault breaker."

By the mid-1980s the Soviet interest in a "military-technical revolution" had risen to the highest military levels. The chief of the Soviet General Staff, Nicolai Ogarkov, became its champion and more openly suggested that the United States had jumped ahead of the Soviet Union thanks to the American military-technical revolution. Ogarkov argued the necessity of increasing the Soviet military budget, shifting decision making on military industrial priorities back to military control, and reorienting procurement toward the kind of electronic investments to which the United States had shown its commitment. Ogarkov's arguments, including an open call for a major budget increase to fund the Soviets' own military-technical revolution, ultimately led to his ouster as defense chief. But by the time Soviet leader Mikhail Gorbachev made Ogarkov a figurehead in the Warsaw Treaty Organization, Ogarkov's articles had stirred enough interest among Russian watchers in the West to reduce his hypotheses of the military-technical revolution to an official acronym, MTR. (A higher form of praise from Pentagon bureaucrats does not exist.)

In his Pentagon Net Assessment office, meanwhile, Andrew Marshall had become convinced that the Soviet military—and the Soviet Union itself—was actually far weaker than the Pentagon realized. The stage was set for a recurring, often nasty debate behind closed doors where military traditionalists insisted on portraying the Soviet military as a global juggernaut capable of exploding across the inter-German frontier, of waging war at sea against the U.S. Navy, and of threatening other strategic areas along its borders.[18] When the Soviet Union finally began to implode in the late 1980s, many in the Pentagon lauded Marshall's assessment as prescient, while he and others who had made that prediction deemed themselves fortunate to have survived after committing the Cold War heresy of identifying Soviet weaknesses rather than issuing alarmist reports about Soviet strength.

It was during this time that one Soviet official made a sour joke that crystallized the dilemma the U.S. military would soon confront. Georgei Arbatov, director of the Soviet Institute of the United States and Canada, a Kremlin think tank, told American reporters in 1989, "You see, we're doing the worst possible thing we can to your military-industrial complex. We're removing your threat and the whole rationale for your military's being." Not everyone in the Pentagon laughed at the jibe. Arbatov was on target in pointing out that the U.S. Defense Department would soon have to readdress basic issues that had been neglected for decades. If the Cold War and the bipolar world defined by the U.S.–Soviet superpower rivalry no longer grounded our national interests and strategy, what did?

Marshall suggested that the best model to study and emulate was the period between the two world wars, which saw considerable military innovation in the United States and elsewhere. Many of the military "revolutions" of World War II—the blitzkrieg, amphibious operations, strategic bombing, the strategic use of aircraft carriers—emerged from the science, technologies, and doctrines of the 1920s and 1930s.

By 1993 Marshall and his analysts had pronounced the phrase "military-technical revolution" inadequate to describe the potential at hand for the post–Cold War U.S. military. Marshall and his colleagues, writing in the pages of U.S. military journals and in internal Pentagon memos between 1989 and 1993, argued that Ogarkov's "military-technical revolution" was too narrow a concept. Based on their studies of earlier periods of military change they suggested that although new technology could significantly alter military capabilities, technology alone was not sufficient to account for the magnitude of change that was now possible. Most of the case studies Marshall had sponsored indicated that big changes in military capabilities took place when new weapons or other military equipment came into use alongside equally pronounced shifts in tactics, doctrine, and military organization. What was involved now was broader than military technology alone.

By 1993 the phrase "military-technical revolution" had been eclipsed in the Pentagon's vocabulary by a newer, more comprehensive term: the American Revolution in Military Affairs.

The new term had important implications given the quandary being debated inside the Defense Department. Marshall concluded a series of studies of previous military revolutions by emphasizing that a true military revolution doesn't happen very often, and that when it does it usually takes time to affect the military and its larger society. Most of the previous revolutions cited in Marshall's studies took decades.

This interpretation fit well with the orthodox view of military change held by the Pentagon leadership at that time. It did not threaten the consensus view of the U.S. military leaders, who, in the early 1990s, strongly favored conserving as much as possible of the Cold War military apparatus rather than changing the U.S. armed forces to meet the challenges and requirements of the new era.

The caution in the early 1990s was understandable—but in the long term, it was a serious mistake. After the end of the Cold War and the U.S. victory in Operation Desert Storm, most U.S. military leaders—including Chairman General Colin Powell and his fellow members of the Joint Chiefs of Staff—knew that we had a first-rate military. So in 1991 the Pentagon was not interested in embracing a Revolution in Military Affairs that called into question many of the premises and assumptions that had for so long defined the size, structure, overseas basing, and missions of the U.S. military. Neither was the White House. So instead of a revolution we had a summer of Desert Storm victory parades, and our domestic agenda focused on economic issues rather than national defense.

DESERT STORM—LEARNING THE WRONG LESSONS

Nearly a decade later, it is difficult to pinpoint the lasting significance of the U.S.-led coalition war against Iraq in the winter of 1991. That the United States was able to organize and mount a major multiservice expeditionary force to the Persian Gulf, under the authority of the United Nations, a force that deterred further Iraqi aggression after its seizure of Kuwait, is clear. Saudi Arabia and the other Gulf emirates remain independent today. That the coalition was able to launch a major air-war campaign followed by a ground-war offensive that ejected the Iraqi Army from the emirate, is beyond debate. Kuwait today is free and prosperous while Iraq remains isolated and poor. (That Saddam Hussein retained his grip on political power in Baghdad following the conflict—and remains there still today despite his isolation from the rest of the world—is also obvious. That Hussein continued as Iraq's leader is a reflection not of allied military incompetence but of the deliberately limited goals of the allied coalition as dictated by the U.N. Security Council resolutions, which explicitly restricted Operation Desert Storm to its goal of liberating Kuwait, and not seizing and occupying Iraq.)

Because of the genuine geopolitical stakes in 1990–91, and because of the size of the competing forces, Operation Desert Storm will be the subject of study and debate well into the twenty-first century, and deservedly so. As one recent defense analysis noted, the Gulf War "profoundly shaped U.S. military thinking throughout the 1990s, placing greater emphasis on precision weaponry, command and control, battlefield surveillance from air and space, logistics modernization, and more extensive use of prepositioned (military) equipment."[19]

For those interested in the future vitality of the U.S. armed services, the question remains: Were the correct lessons learned, and were the right questions asked?

The Gulf War set the stage for the United States to win and sustain its deserved reputation as the world's preeminent post–Cold War military power. Our role in the conflict inspired a national pride at home that we had not seen since World War II. To then President George Bush and many of us in the senior military leadership at that time, the desert victory helped purge us of the "Vietnam syndrome," a national reluctance to become engaged in a major military operation out of fear it could bog down. Rick Atkinson, profiling the generals and admirals who directed the Desert Storm campaign as veterans of our failed war in Southeast Asia, caught the mood of vindication as they met at the village of Safwan in southern Iraq to impose a cease-fire on the Iraqis:

"They had stayed the course after Vietnam, vowing to restore honor and competence to the American profession of arms and, most important, to renew the bond between the republic and its soldiery. This—Safwan, March 3, 1991—was their vindication. For Norman Schwarzkopf and his lieutenants, this war had not lasted six weeks, but twenty years."[20]

The world correctly recognizes Operation Desert Storm as the sine qua non of military prowess. Our young men and women had crushed the world's third largest army. And they did it the way Americans always had hoped it could be done, with dispatch and skill.

New Technology and Weapons

Despite panicked predictions by some that the United States and its allies might suffer as many as 50,000 casualties in the desert, we lost only 293 men and women, and half of them outside of combat. Although every death in war is tragic, the numbers were incredibly low given the inherent capabilities of the Iraqi Army. Despite predictions that the United States might see hundreds of its combat aircraft shot down, only 22 American and 9 allied aircraft out of an armada of over 1,600 combat aircraft were lost in combat (another 8 aircraft went down in noncombat incidents), another astoundingly low figure. No U.S. ships of the nearly 300 that participated in Desert Shield and Desert Storm were sunk, and only two were damaged by Iraqi mines. As the statistics from the postconflict assessments rolled into Washington, it seemed increasingly clear inside the Pentagon just how remarkable our victory had been. We had moved more tonnage, over greater distances, in less time than we had in the Normandy invasion in World War II. We had delivered more ordnance by air in the six weeks of Desert Storm than in the "most productive" six months of World War II. In the maneuver that we called the "left hook," the U.S. Army swung its M1A1 Abrams tanks and M2 Bradley fighting vehicles over longer distances in less time than any army in history, achieving a rate of advance nearly twice as high as the majority of other twentieth-century military routs.[21] Indeed, the 58-ton Abrams tanks, long derided by critics as prone to mechanical flaws, outmaneuvered and outfought the Iraqis because of the tanks' superior design, better-trained crews, and high-technology gun systems, which enabled American tankers to locate enemy units and destroy them long before Iraqi sensors even knew an attack was under way. The Navy conducted the largest, most effective blockade in history, using its surface warships and land-based patrol planes as a coordinated team. The U.S. Marine Corps killed more enemy soldiers more quickly than at any time in the Corps's history. U.S. Navy and Air Force aircraft quickly swept the Iraqi Air Force from the sky, dismantled the Iraqi air defense system, and turned out the lights in Baghdad. And we had proof of all of it, from the videotapes that recorded the precision air strikes, to the reports of the Iraqi prisoners, to the data that poured in from the damage assessments.

Many experts have argued that the Persian Gulf War itself constituted a revolution in military affairs—at least in terms of its being a massive firepower demonstration utilizing the products of a technological and strategic transformation. Perhaps it is because Desert Storm was the first major conflict played out live on television that so many people held that viewpoint. In any case, they are incorrect.

True, there were dramatic and exciting moments to watch on television and (as a senior military commander) to learn over our encrypted communications network. We reveled in an impressive demonstration of weapons never before tested in combat.

As the 6th Fleet commander, in early 1991, I ordered the USS *Pittsburgh*, a Los Angeles–class attack submarine operating under my command in the Mediterranean, to initiate Operation Desert Storm by launching the first Tomahawk land attack cruise missile ever fired in anger. Built on U.S. technology, the unmanned, turbojet-powered missile flew hundreds of miles on its own on a preprogrammed route, slipped unobserved into northern Iraq and on to the area around Baghdad, then located, homed in on, and burst upon its target.

Nor was this the exception. A lengthy roster of untried weapons were fired in anger for the first time in Operation Desert Storm, and all performed excellently. The F-117A Nighthawk stealth fighter (which had made a one-flight sneak appearance in Panama two years earlier) did the yeoman's share of high-priority pinpoint bombing over Baghdad and other target areas heavily defended by Iraqi antiaircraft defenses. The AH-64 Apache helicopter gunships, armed with Hellfire antitank missiles, played key roles both in defense of the coalition ground force and as a long-range attack air weapon capable of ravaging the enemy far from the front lines. The Army unveiled its nonnuclear (Army Tactical Missile System (ATACMS), which provided an extended-range artillery capability, fired from the Multiple-Launch Rocket System. The Navy and Air Force both operated new Unmanned Aerial Vehicles (UAVs)—small pilotless drones used for surveillance and artillery spotting—and scored a historical "first" when one group of Iraqi soldiers actually attempted to surrender to a UAV flying overhead. The experimental Joint Surveillance and Targeting Radar System (JSTARS) and mature Airborne Warning and Control System (AWACS)—both installed on converted Boeing 707 airframes—provided the air and ground commanders for the first time a radically comprehensive picture of the airspace and ground terrain, enabling allied air units to operate more safely in an area where Iraqi missile and antiaircraft units still posed a potential threat. Operation Desert Storm was also the first major conflict in which the U.S. Space Command unveiled its long list of capabilities, including global communications, weather support, reconnaissance, and missile-launch warnings. The worldwide Global Positioning System satellite network provided both aircrews and ground units with accurate navigational fixes, minimizing the chances that they would blunder into accidental combat with allied units or become lost. On the ground, individual soldiers benefited from a new class of night-vision devices and digital radios (difficult to intercept and all but impossible to jam). All of these systems—engineered and acquired in the late 1970s through the late 1980s—made the allied victory inevitable and our historically small loss of life probable.

No Advances in Cooperation Among the Services

On the surface, the Persian Gulf War also broke new ground in terms of the success of interservice cooperation and "joint" operations between the different U.S. armed services. Operation Desert Storm seemed to reverse a long and sad streak of military operations since Vietnam in which miscommunication, rivalries, and mismatched combat doctrine had led to disaster, such as the botched attempt by a "joint" task force to rescue our embassy hostages in Iran in 1980, and the inefficient and poorly administered intervention in Grenada three years later.

The Persian Gulf conflict was the first major U.S. use of military force under the Goldwater-Nichols Defense Reorganization Act of 1986, which had enacted serious reforms aimed at breaking through the ancient rivalries and bureaucratic skirmishes that for so long had crippled our armed services. The law centralized overall military authority in the chairman of the Joint Chiefs of Staff, empowered the regional commanders in chief (such as General Norman Schwarzkopf) to determine how military operations were to be organized, and carried out (without political pressure from the Pentagon)[22] and for the first time in U.S. military history rewrote the personnel manuals and career policies to make serving in a joint duty position—a staff or command billet formally identified as supporting a multiservice headquarters—not only a positive step in terms of promotion considerations, but essential for ascending to senior command.

The national command structure on the surface did function smoothly, each part playing its appointed role with Schwarzkopf in charge of prosecuting the war. Under him were five subordinate commanders, all three-star generals or admirals, who led the "component" commands.[23]

Schwarzkopf himself would recall in his postwar memoirs that just seven years before the Persian Gulf War he had been snatched from his post as commander of the 24th Mechanized Infantry Division in Georgia to become the deputy commander of the U.S. task force that invaded Grenada in October 1983. Reporting to the naval task force, Schwarzkopf learned that he had no formal authority to do anything except "advise" the naval commander, Vice Admiral Joseph Metcalf. There was no formal military arrangement establishing Schwarzkopf's role as deputy, he had no legal authority within the Atlantic Command (then dominated by the Navy), and he found himself rebuffed when he ordered a Marine Corps colonel to fly Marine Corps helicopters from the helicopter carrier USS *Guam* to pick up Army Rangers for a rescue mission. After the first day of operations had gone badly, Metcalf, to his credit, turned to Schwarzkopf and said, "I will confess that I know very little about ground operations. Would you make up the plans for tomorrow and write the orders that we should give to the forces?"[24] In that sense there had been progress. When Schwarzkopf arrived in Saudi Arabia in late 1990 there were at least clear lines of authority and a team of military forces from the four U.S. combat services (as well as the coalition) that had trained for operations in the region under his command.

But many flaws remained—flaws not from poor performance, but from an ingrained command hierarchy and an outmoded concept of war that had taken root during World War II and then during the Cold War. Desert Storm was a joint military operation in name rather than in fact. All the military services were represented and participated, but the way they performed was essentially a mirror image of how they had been operating for the past thirty years. The battlefield was divided among the service components. Lines were drawn, if not in the sand, certainly on the maps that laid out areas of responsibility. Yet these were more than areas of responsibility; they were areas of service discretion, in effect fiefdoms where, within general guidelines, each service component ruled.

The fiefdoms existed not only because of tradition, service rivalry, and the egos of the commanders; they were also there because of technical limitations. We did not have the communications capability to do it differently. The Army could communicate internally with itself quite well. It had integrated the right frequencies, protocols, ranges, experience, and doctrine to do so. It was much harder for the Army to communicate with the U.S. Marine Corps. The Marines on the ground communicated quite well with their fellow Marines in the air. But they had a lot more difficulty sending messages to and receiving messages from the Air Force's jets and the Army's attack helicopters.

These technical difficulties showed up in the operations. In the ground conflict, the Marines focused on a particular objective—the liberation of Kuwait City—without the benefit of Army artillery or the support of the Army's new ATACMS ballistic missile. The Army swept to the west into the vast Iraqi desert without the benefit of Marine Corps close air support aircraft such as the AV-8B Harrier fighter, which can take off and land vertically without requiring a runway.

The air campaign that America saw as a seamless flow of precision power was actually an effective edifice of separate building blocks. True, it was orchestrated by daily "air tasking orders," detailed lists of targets to be struck and air units assigned to carry out the missions, all of which came from Central Command headquarters. Prompt execution of those orders was another painful issue. Once devised in Riyadh, the tasking order took hours to get to the Navy's six aircraft carriers—because the Navy had failed years earlier to procure the proper communications gear that would have connected the Navy with its Air Force counterparts, allowing the massive data to be sent electronically and instantly. To compensate for the lack of communications capability, the Navy was forced to fly a daily cargo mission from the Persian Gulf and Red Sea to Riyadh in order to pick up a computer printout of the air mission tasking order, then fly back to the carriers, run photocopy machines at full tilt, and distribute the document to the air wing squadrons that were planning the next strike. As the air campaign continued, the Navy moved away from the main air campaign, opting instead to cobble up missions that gave them more direct control over their aircraft. The Marines—who have always jealously guarded their small air arm as a close air support weapon—simply exempted their aircraft from the Central Command's air tasking order altogether, arguing that, after all, the Harriers and helicopter gunships were on hand exclusively to support Marine Corps ground operations. This was just fine so far as the Air Force was concerned, because that way the Air Force could run its operations with less intervention by the Navy or the Marine Corps! As the conflict went on, the issues that divided the different military services grew stronger, and the impetus for cooperation became more and more strained.

The management of the air operations was identical to the flawed and rigid service policies of the past thirty years. Air Force and Navy aircraft hit many of the same targets. But they did not actually operate together—because they couldn't. Navy carrier warplanes did not have the same type of "Identification, Friend or Foe" transponder (a device that sends a coded pulse back to another aircraft or radar receiver confirming its identity) that the U.S. Air Force had, so to prevent accidental shootdowns, Navy bombers had to be kept as far away from Air Force fighters as possible. As a result, the U.S.-led air campaign was deprived of opportunities to create a joint force strike team. Instead, all–Air

Force and all-Navy attacks were the norm, with each service performing support missions such as surveillance, local air defense suppression, bombing, and post-strike bomb-damage assessment. Although in most cases this division of labor worked adequately, it was not the best use of the aircraft and weapons, because both services' strengths could have been strengthened even further had the services coordinated their communications, training, and preparation for combat.

For example, the Navy EA-6B Prowlers were regarded as the most superior electronic warfare planes in the entire U.S. military (the Air Force used a variant of the F-111 to carry out its missions to electronically jam enemy radars). On the other hand, the Air Force F-15E Strike Eagle was vastly superior to the A-6E Intruder in bombing with laser-guided bombs. Put the Navy jammer and Air Force strike bomber together in the same mission and—wow! But even in 1991 the idea of mixing combat aircraft from different services in a single strike remained as heretical as it had been in Vietnam a generation earlier, when the Air Force and Navy arbitrarily divided North Vietnam airspace into "route packages" designed not to defeat the enemy, but to keep U.S. Air Force and U.S. Navy aircrews out of each other's hair.

Still, we destroyed the Iraqi Air Force with dispatch. The Iraqi aircraft that survived did so only by hiding or fleeing to Iran, and there were no successful Iraqi air attacks on coalition facilities.

But there was one close shave, and it illustrated how little our thinking had changed in the previous two decades. Two Iraqi Mirage F-1 fighter-bombers almost made it to within striking distance of the logistics hub at Dhahran on Saudi Arabia's eastern coast by streaking southward along the coast. They flew along the seam of our air defenses where, as tradition dictated, we divided responsibilities between the U.S. Air Force and carrier-based Navy aircraft. The seam did not exist because of a flaw in our surveillance capability or technical limitations in our weapons. Both the Air Force AWACS plane, orbiting over Saudi Arabia, and the Navy's Aegis system carried by ships in the Persian Gulf identified and tracked the incoming aircraft. Shorelines, however, had always been the demarcation lines between our land and sea forces, and this conceptual residue from the past now opened a pathway for our opponent.

The Iraqis didn't pull off the mission. We reacted just in time, helping to vector a Saudi F-15C interceptor to the kill. The Saudi Air Force had a spectacular success and morale was boosted in the multinational coalition. But after the congratulations and feelings of relief, disturbing questions were not asked. Why had this happened? And what if the Iraqi pilots had gotten just a little bit closer to Dhahran, close enough to launch their weapons?

Another critical aspect of Operation Desert Storm was the logistics effort that moved so much materiel into the theater. In many respects, the effort was masterful. But much of what we brought in piled up on the Saudi docks and holding areas anonymously. We lost track of what was in the containers, many of which came from Germany containing "sailboat fuel," as one wry soldier had labeled it: a container containing nothing but air that had been hastily thrown into the outbound flow of cargo because there wasn't enough time either to fill it with supplies or to haul it out of the way. Because it was too hard to search through the growing mass of supplies, many units simply reordered what they needed. The piles of equipment, spare parts, and supplies kept growing throughout the conflict, all the time becoming an increasingly more lucrative target. By the time the war was over, we had piled up roughly twice as much ordnance as we used throughout the shooting. Some of it is still there.

The False Revolution

It's hard to argue with success, and our successful ejection of the Iraqi Army from Kuwait was a rare, unambiguous victory on the ground, at sea, and in the air. *But our military performance did not constitute a true military revolution.* Here's why.

First, the underlying U.S. response to Iraq's aggression fell under the timed-phased-deployment formula that has governed our major military campaigns since World War II. Upon receiving the news that Saddam Hussein had sent his army over the Kuwaiti border, the Bush Administration ordered the Pentagon to initiate its operational plans for the defense of Saudi Arabia. Operation Plan 90-1002 was a carefully organized, complex roadmap for the movement of U.S. forces from bases in the continental United States and Europe to the Persian Gulf region. The principles underscoring the plan were (1) the *sequential* deployment of U.S. air and ground units to Saudi Arabia and other friendly Gulf nations, (2) the organization of a *defensive* plan to deter and prevent Iraq from attacking across the Kuwaiti-Saudi border, (3) a *gradual* buildup of forces in the region to consolidate the allied military position, and, finally, (4) the drafting and execution of an air and ground campaign to expel Iraqi forces and liberate Kuwait.

Second, the underlying premise of the organization, training, and deployment of this force was the need to accumulate an overwhelming *mass* of military power—strength and numbers—to carry out the operation against Iraq. This is the model—the "template"—that has governed U.S. military contingency planning since the Japanese attack on Pearl Harbor nearly six decades ago. As defense analyst Jeffrey Cooper recently wrote, the "American way of war" for most of this century has followed identical assumptions:

> Most current thinking in the United States about the future conduct of warfare and the utility of air and space power occurs within a context of powerfully constraining but usually unstated assumptions . . . deriving from more than 50 years of experience and habit dating back to before World War II. Perhaps the most important of these factors is the emphasis on overwhelming force and joint (usually meaning four-service) operations, and the phased, sequential nature of our operational plans—factors deeply imbedded in our historical images of war from Gen. Dwight D. Eisenhower to Gen. H. Norman Schwarzkopf. Although understandable in the circumstances of the Cold War, this context poses significant problems for U.S. military policy in the new era.[25]

Cooper's assessment captures a fascinating contradiction of the Persian Gulf War. From the outset, Air Force planners—notably strategist Air Force Col. John Warden—designed an air campaign that would wage "parallel war" against Iraq in the form of simultaneous strikes against enemy command-and-control centers, air defenses, and other military targets. As Air Force historian Richard P. Hallion later observed, the air war would be "quick, overwhelming and decisive." Cooper noted the staging of the allied military coalition—including the deployment of combat aircraft—mirrored the World War II–era expeditionary campaign. And the U.S. armed services were still unable to combine their different strengths into a truly unified military force. The opportunities for expanding combat power through the synergy created by joining the disparate strengths of the different services were squandered. Part of the reason was technological—for instance, the mismatch between Navy and Air Force communications. Part of the reason was cultural; for instance, relying on its long-term tradition of conducting independent operations, the Navy did not pursue significant programs that could have helped its forces coordinate combat operations with the other services. But the largest reason, in my view, is that our senior military commanders—like the general public, the congressional overseers, and the civilian leaders in the White House and Pentagon in 1991—simply failed to grasp the implications of the new information technology, that it made possible a vast and unprecedented expansion of American combat capability.

In that sense, Operation Desert Storm was a transitional conflict that contained the seeds of a revolution but did not constitute the transforming event itself. A decade after the Gulf conflict, that revolution has still yet to occur.

CHAPTER 3

THE TECHNOLOGICAL BASE

The wide array of hardware and software currently adaptable for military use, and even more powerful applications we can foresee in the near term, will empower the U.S. military to maintain combat superiority over all other nations well into the twenty-first century.

Technology alone is not enough, however. The Revolution in Military Affairs will remain only a promise unless the Pentagon correctly rewrites combat doctrine to forge truly joint operations, reorganizes leadership and training programs, and fully incorporates the new technology throughout the four combat services. And even so, there is no guarantee that the amazing potential of this military revolution will not be misused and squandered by civilian leaders who adhere to unrealistic policies governing the use of military force.

But the technological base of the Revolution in Military Affairs remains the central component of a transformed twenty-first-century American fighting force and the best hope for the United States to keep its armed forces superior to any other nation's. There is nothing mysterious about how this revolution will work. A U.S. military commander in the new century, like his predecessors down through history, will still prepare for battle by collecting information relevant to his mission, processing it to extract what is militarily significant, and communicating that subset of data to his combat forces. Combat units will use that information to locate the enemy, outmaneuver its combat units, and strike targets across the length and width of the battlefield. Simultaneously, the U.S. commander will use the information in hand to respond quickly to any enemy countermoves, to deny the enemy commander the ability to maneuver to threaten the force, and to protect the U.S. troops and support units. The technology underpinning the revolution will enable U.S. forces to carry out military operations better, faster, and more effectively than any foe.

The key difference is the order of magnitude of the technological change. The quantity and volume of information that will be handled will dwarf what was available in any previous conflict. The speed at which the information will be transmitted, received, and processed will be unprecedented, and the information itself will tell us more about the enemy and the battlefield than any previous commander could have imagined.

THE SYSTEM OF SYSTEMS

What kind of technology is central to the Revolution in Military Affairs? Computers, sensors, satellites, and wireless communications form the backbone of this technological transformation. One can envision an infantry squad on a battlefield several decades from now preparing to engage an enemy force. As defense journalist James Adams has projected:

> In this new world the soldier will be the young geek in uniform who can insert a (computer) virus into Teheran's electricity supply to plunge the city into darkness. . . . The soldier will also be the man or woman equipped with a uniform powered by body heat that automatically adjusts to the environment and that relays location and vital signs back to base. That soldier will have on his head a helmet that allows him to see in all conditions, to locate incoming fire and return it with deadly accuracy, and an eyepiece that will provide his location, the location of his enemy and the locations of others in his patrol. . . . He will be equipped with tiny airplanes no larger than a small notebook that will fly ahead and show him the terrain and the enemy.[1]

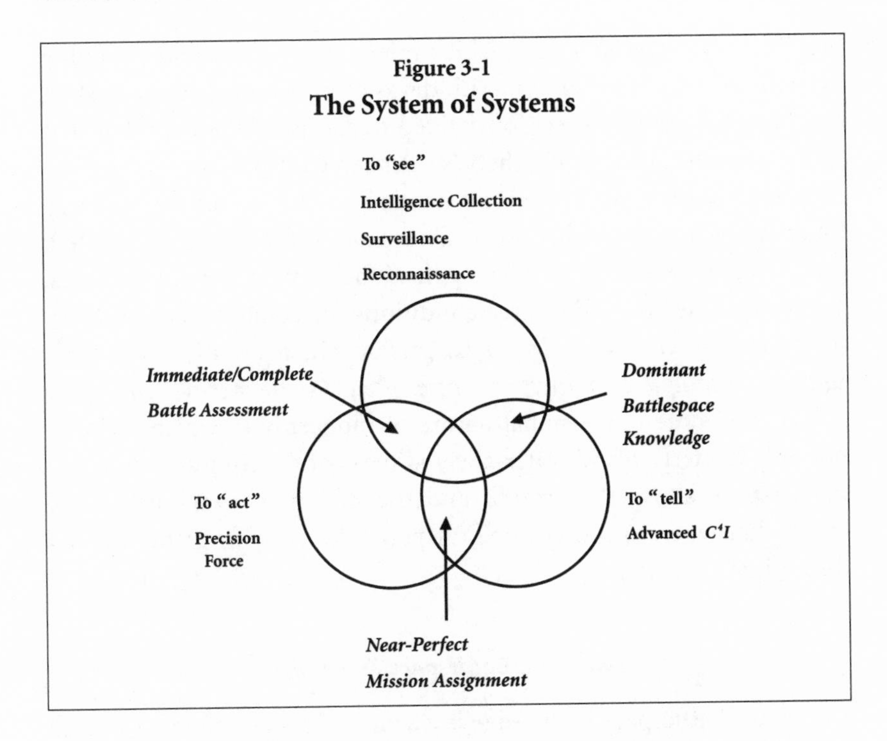

Because the information technology driving the current transformation involves the interaction of a number of separate technical applications—materials sciences, cybernetics, and electronics, for example—we use the notion of a broad *system of systems* to illustrate the technical changes inside the U.S. military. The Pentagon has embraced this term, which is depicted in Figure 3-1.

Figure 3-1 shows how different technological applications will allow the U.S. military commander to carry out his traditional functions to "see," to "tell," and to "act." The first area (to "see") depicts the arena of *intelligence collection, surveillance,* and battlefield *reconnaissance* that provides the commander the means to collect information about the enemy and the area in which the opposing forces will contend. The second area (to "tell") embraces advanced *command, control, communications, computers, and intelligence*—known by its military acronym of C^4I—which enables the commander to transfer data and communicate quickly and effectively with his subordinate commanders even though they are dispersed across thousands of square miles. The third area (to "act") involves *precision force,* which refers to the commander's ability to bring military force to bear with devastating accuracy using existing and planned types of precision-guided weapons, of the type that first captured public attention in the 1991 Persian Gulf War.

When technology is correctly applied to the traditional military functions—to see, to tell, and to act—a powerful synergy is created, producing an effect much greater than the sum of the components. Together, these create the three conditions for combat victory: *dominant battlespace knowledge, near-perfect mission assignment,* and *immediate/complete battlespace assessment.* (The term *battlespace* reflects the three-dimensional nature of modern war, which includes not only the terrain and water-covered parts of the contested area, but the airspace above the ground and the region of near-earth space above that where orbiting satellites play a key role in gathering and disseminating data.)

Dominant Battlespace Knowledge

Dominant battlespace knowledge is a senior U.S. commander's overall comprehension of the enemy, his own forces, the battlefield terrain, and any other factors that will influence the course of battle—such as the ability to keep the enemy from knowing what he knows. It rests on advanced sensing and reporting technologies and includes both the platforms and the sensors we associate with intelligence gathering, surveillance, and reconnaissance—and reporting systems that provide better awareness of our own forces, including the flow of logistics, the location and activity of combat units, and even the status of civilian noncombatants. Other elements include the commander's awareness of external factors that may affect the fight such as weather conditions, landscape features that will affect the movement and safety of forces, or electromagnetic conditions such as enemy radio jamming.

Let's take as a hypothetical example a U.S. Army corps commander whose forces are arrayed against the North Korean Army in the event of an invasion of South Korea. His battlefield headquarters would be a nerve center linked to space-based satellites, ground- and sea-based sensors, and manned surveillance aircraft continuously scanning the mountainous terrain of the Korean peninsula to detect enemy troop movements, to identify specific threats facing his deployed forces, and to provide him with a constant, real-time image of the battlespace embracing the terrain and airspace above it. That data as well as other intelligence (such as intercepted enemy communications) would provide the commander the ability to direct his ground and air forces to attack the enemy in such a way as to thwart its drive to overtake U.S. and allied forces defending South Korea. At the same time, secure high-speed communications between his command center and all units under his control would enable the commander to manage the flow of supplies and ammunition to fighting units in such a way as to maximize their movements against the enemy, while protecting the logistics flow from enemy disruption.

In order to achieve dominant battlespace knowledge, our military must have both the equipment and the people who know how to use it. First, the U.S. commander must have the ability to collect data and information on the enemy through a full spectrum of means ranging from special operations surveillance teams to satellite sensors. Equally important, the commander must have the personnel and equipment in place to collect the massive influx of data and process it into relevant knowledge.

Near-Perfect Mission Assignment

Near-perfect mission assignment comes into play once the data is processed into information that is pertinent to the combat mission, and dispatched by the commander to his field units. This information may consist of reports summarizing the location and strengths of enemy forces and their current and projected direction of movement; it may include updated terrain maps of the battlefield showing not only troop locations but updated markers showing which roads and bridges have been destroyed in previous fighting; it may show tactical battlefield displays indicating the actual movement of tanks and infantry fighting vehicles—information gathered by a host of intelligence-gathering systems such as unmanned aerial vehicles, satellites, and infantry soldiers—and projected by computers on a digital map of the battlefield, with special visual display codes to differentiate between "friendly" and enemy vehicles. The exchange of information between headquarters and fighting units, and the transformation of raw intelligence data into relevant information, is made possible by high-volume, high-speed secure computer networks and communications channels linking all elements of the Army corps into a single entity. Subordinate combat commanders use the available information to target and organize the battle to destroy the most important parts of the enemy force using the best-suited weapons available to carry out the mission.

Moving from blindness to total vision on the battlefield is one of the key aspects of the Revolution in Military Affairs. Even in the air campaign in Kosovo in the spring of 1999, incomplete and poorly organized information networks meant that there was still a delay between the time when an unmanned aerial vehicle used for reconnaissance located a target, such as a Serbian tank formation, and the time when a U.S. or allied fighter could reach the target area to bomb it. One of the key premises of the military revolution will be to attain a continuous, real-time exchange of targeting data to permit the fastest possible response with combat aircraft or artillery fire.

Immediate/Complete Battle Assessment

It is not enough for the commander, his subordinates, or soldiers in the field to possess information on the enemy's location and to know which of their weapons should be assigned for the attack. The U.S. commander and his field units must have the ability to assess immediately the success or failure of the strike so that they can carry out new attacks if necessary. Today assessment is accomplished through post-strike photography using manned aircraft or photo-reconnaissance satellites. Under the Revolution in Military Affairs, the Pentagon would deploy a much larger number of manned aircraft and unmanned aerial vehicles than it has thus far in order to provide saturation coverage of target areas in the shortest possible time, while pressing ahead with new surveillance capabilities employing infrared imaging and space-based radar to obtain accurate pictures of the bombing even through cloud layers or smoke. The rapid and successful transmission of this imagery to the correct assessment team requires a third area of improvement: the full integration of intelligence, surveillance, and reconnaissance with the weapons—the fighting force itself—under a commander's supervision.

LESSONS FROM NORMANDY, LESSONS FROM NASIRIYAH

Military commanders throughout history have striven toward the conditions or "capabilities" for combat victory, but for most of the twentieth century, including World Wars I and II, Korea, and Vietnam, they were regarded as ideals to strive for but impossible to attain. During the 1990s, in military operations from Operation Desert Storm to Kosovo, the U.S. military acquired and employed a number of sophisticated command systems and smart weapons that enabled U.S. commanders to apply information-age technology in battle: In the Kosovo air campaign in the spring of 1999, the United States was able to fire several hundred Tomahawk cruise missiles and air-drop Joint Direct Attack Munitions with pinpoint targeting accuracy despite clouds or weather blocking the target, thanks to an onboard computer on the weapons that made use of the Global Positioning System satellite network to refine their position against target coordinates. The use of such weapons provides a tantalizing glimpse of how effective the "system of systems" could be. On the other hand, the U.S. forces in the NATO air campaign against Serbia still did not have the ability to transmit such targeting information directly from surveillance aircraft to fighters—a serious lapse. (See Chapter 5 for discussion of Kosovo conflict.)

Still, we have come a very long way since the battles of World War I, in which thousands of young soldiers perished as cannon and machine-gun fodder in the battles of attrition fought in the trenches of Europe. The distance we have come can be charted by two combat milestones of the twentieth century—the allied invasion of Normandy on June 6, 1944, and the "left hook" allied ground offensive against Iraq in 1991. These two battles are instructive in showing both the great gains that have been made in military technology and the military's failure to take full advantage of them—with grievous consequences.

Normandy: The Commanders Were Blind

The largest amphibious landing ever planned in world history was Operation Overlord, the Allied invasion of France on June 6, 1944. On the first day the Allied generals planned to land over 154,000 soldiers on the Cotentin Peninsula of northwestern France, with 50,000 troops coming ashore in the first wave alone. The U.S.-led alliance had 47 Allied divisions and 620,000 soldiers overall ready to take on 60 German Army divisions in France and the Low Countries.

To carry this off, the Allies mustered 4,400 warships, transports, and landing craft to carry the bulk of the force from the sea to the beach (20,000 American and British airborne soldiers were dropped behind German lines in the early hours of D-Day to disrupt enemy communications and seize strategic bridges to prevent German reinforcements from arriving in time to crush the landings).[2]

The design of the invasion plan, the invasion itself, and subsequent Allied efforts to "break out" of the German lines were governed by a number of bleak realities confronting the Allied commanders under General Dwight D. Eisenhower. These were the realities of combat in the mid-1940s: primitive surveillance capabilities that generally failed to inform the commanders of battlefield events; fragile communications technology on which the commanders relied to direct the planned operation; and the age-old reliance on mass—overwhelming numbers of ships, aircraft, and soldiers—to compensate for the many unknowns of battle.

Historical accounts of the Normandy invasion provide a detailed chronology of bravery, improvisation, chaos, and violence all predetermined by gaps in the Allied command's knowledge about the Normandy beaches and the German defenders. In the months before the invasion, Eisenhower and his lieutenants made their best efforts to compile an accurate picture of the geography and terrain of the Cotentin and the location and strengths of the German army defenders dug in at the beachhead. As the Allied soldiers learned immediately on Omaha Beach, at the cliffs of Pointe du Hoc, and at the Douve River bridges, the intelligence was mostly wrong, and critical battlefield information was frequently incomplete and inaccurate. The Normandy invasion would become a classic attrition battle where the gap between pre-combat estimates and the "ground truth" of the invasion would be filled with the bodies of dead American and Allied soldiers. It's instructive to review Operation Overlord using the contemporary terms of the current Revolution in Military Affairs.

First, neither Eisenhower nor the Germans possessed sufficient information about the opponent to achieve *dominant battlespace knowledge.* Allied intelligence-gathering efforts prior to the invasion were extensive, including spying conducted by members of the French Resistance underground, Allied aerial photography, and even last-minute reconnoitering of the Normandy beaches by frogmen. Still, the overall picture of the battlefield on the eve of the invasion was still far from complete and, in several instances, fatally inaccurate. Germany, on the other hand, had an even poorer grasp of the Allies' strengths and intentions, thanks to the inability of the German Air Force in 1944 to carry out photo-reconnaissance missions over Allied staging bases in southern England. And a masterful Allied deception effort called "Plan Bodyguard"—employing captured German spies who fed false information back to the German high command—lulled Berlin into believing that the actual invasion would occur along the northwest French coastline of Pas de Calais later in the summer of 1944 instead of at Normandy. The deception effort included the creation of fictitious Allied commands with agents broadcasting messages meant to be intercepted by German eavesdroppers telling of the assembling of a massive ground force for the invasion of Norway (as a result the Germans were prompted to pin down their own forces there rather than transfer them to France).

Neither the Allies nor the Germans could have adequately predicted the weather as we can today using photo satellites dedicated to tracking storm systems. With the 1940s-era weather prediction technology consisting of radio messages from distant tracking stations and ground-based thermometers and air pressure barometers, meteorologists could only guess at weather conditions twenty-four or forty-eight hours into the future. Because of unusually stormy weather conditions in the English Channel, the invasion originally set for June 4 was postponed, and subsequently Eisenhower's weather officer could give the supreme allied commander only an educated guess that the storms would abate on June 6. Faced with the prospect of having to postpone the invasion again until June 19 (to await the return of favorable tide conditions) Eisenhower had to make a guess—take a gamble—th
ended in a massive military failure with paratro
into enemy strongpoints by unforeseen winds and
phibious landing craft foundering or sinking in sto

before they reached the shore. That Ike gambled correctly is known to history.

An assessment of Normandy in terms of near-perfect *mission assignment* (particularly the selection and deployment of weapons against particular targets) produces a dreary catalogue of fog and friction in war. Hundreds of warships, from battleships to destroyer escorts, pounded the Normandy coastline prior to the order to send in the landing craft. But naval optical gunsights weren't powerful enough to gauge the success of the bombardment. (It wasn't successful: the Germans were too deeply dug in, and most of the shells fell too far inland.) Nor was a massive aerial attack of over 14,000 sorties (individual missions) by over 5,000 heavy and medium bombers and another 5,000 fighters out of England any more successful against the beach defenses. The aircrafts' missions were designed to overlap with the naval bombardment, but the aviators fared even worse, thanks to the inherent inaccuracy of Allied bombsights and the presence of a thick cloud layer below that World War II-era technology could not penetrate. The naval shells and bombs plowed up the Norman countryside and missed the German beach defenses altogether.

The commanders' intelligence gathering was also spotty. In one example, Allied intelligence misidentified the German units defending Omaha Beach—the landing zone for the U.S. Army V Corps, and the 1st and 29th Infantry Divisions—with fatal consequences. Eisenhower's intelligence specialists believed that only one battalion (about 750 men) of a second-rate army unit manned by Poles and turncoat Russians was on guard duty at Omaha Beach. In fact, the combat-hardened 352nd Division of the German Army had three battalions—more than 2,000 men—dug in along the beach, with a network of pillboxes linked by deep trenches. The German rifle and artillery fire massacred the first wave of American soldiers; several units saw over two-thirds of their fellow soldiers killed or injured as they came out of their landing craft.

Other intelligence failures litter the histories of D-Day. Several miles inland, aerial photographs showed an area of flat farm grassland that appeared to be an ideal paratrooper drop zone. The photographs failed to alert Allied planners that the "grassland" had been intentionally flooded by the Germans (the water did not appear on the photos), and scores of American airborne soldiers drowned on landing.

The heroic scaling of the cliffs at Pointe du H
of the 2nd Ranger Battalion as German gunn
into the climbing soldiers—one of the most he
invasion—was a mission ordered up by faul
tion: the heavy artillery pieces that were the
replaced by dummy guns made out of tele
photo-intelligence had missed the trick (the Rangers
destroyed the guns about 1 kilometer inland).

In order to translate knowledge into effective combat performance, a military leader must exercise assured command and control of the forces under him. At Normandy, senior U.S. and British generals remaining behind on the warships as their lead echelons hit the beach were immediately and completely cut off from the fight. Again, the communications technology available to World War II units—radio sets based on fragile electron tubes with inadequate transmitting power—was both too primitive to work in the chaos of the massed battle and too fragile to withstand the inevitable physical shock of the battlefield. The first indication that there was a breakdown in communications occurred during the night parachute and glider landings, when the massed armada of transport planes carrying 13,400 American and 7,000 British paratroopers entered French airspace and came under German antiaircraft fire. The tightly packed aircraft formations fell apart in confusion at the moment the paratroopers were poised to jump. As one historian noted, "By 0400, the American paratroopers and gliderborne troops were scattered to hell and gone across the Cotentin."[3] Once on the ground, the paratroop commanders found they had lost all contact with the bulk of their soldiers owing to the scattering of the men and the poor state of their communications gear. Military historian Stephen Ambrose wrote:

> There was virtually no overall control (of the U.S. paratroopers) because it was impossible for the generals and colonels to give orders to units that had not yet formed up. The groups that had come together were unaware of where they were or where other groups were, a problem that was greatly compounded by the ubiquitous hedgerows.
>
> Radio communication could have overcome that problem, but

> radios had been damaged or lost in the drop, and those that re working were inadequate. The SCR-300 (Signal Corps Radio 300), which weighed 32 pounds, had a speaking range of five miles, but only under perfect conditions. The more common SCR-536 (walkie-talkie) . . . had a range of less than one mile.[4]

On the invasion beaches, the same chaos prevailed. A combination of several factors—the inadequate pre-landing bombardment, the deeply dug-in German defenses, and rough tides that scattered the landing craft far from preassigned target beaches—severed the Allied command from its front-line units. Years after the invasion, General Omar Bradley, who as commander of the First U.S. Army was responsible for the advance American force of two Army corps headquarters and three divisions assigned to Omaha and Utah Beaches, would recall his feeling of utter helplessness once the battle had begun. "Omaha Beach was a nightmare. Even now it brings pain to recall what happened there on June 6, 1944," Bradley wrote in his memoirs. He described the day as "a time of grave personal anxiety and frustration" because of the breakdown in communications with the forces ashore that left him relying on only fragmented messages from staff officers who rode in on landing craft and attempted to discern the course of the battle through quick glimpses from several hundred yards offshore. Bradley continued, "I gained the impression that our forces had suffered an irreversible catastrophe, that there was little hope we could force the beach. Privately, I considered evacuating the beachhead."[5] Ambrose tersely recounts that Bradley and the other commanders were "helpless observers" held hostage to the prearranged operational plan because they were incapable of receiving timely intelligence about the course of the battle, nor could they have communicated any orders to the forces ashore to change the plan to capitalize on the fluid battlefield situation.

In explaining the Allied victory, historians who have long studied D-Day in Normandy have cited several factors they believe compensated for the loss of commanders' battlefield awareness, the poor intelligence, the breakdown in communications, and the frequently inadequate performance of weapons. First, the Germans were equally confused and were hamstrung in responding to the invasion because of orders from Berlin to keep reserve tank divisions inland (Hitler still believed the real strike would come at Pas de Calais). Second, the Allied soldiers in numerous instances brilliantly improvised when the original plan or mission fell apart. For example, when naval commanders learned that the planned shore bombardment had failed to suppress the German defenders, a handful of Navy destroyers risked running aground by steaming right up to the beach and blasting the defenses at point-blank range.

But the main reason for the Allied victory at Normandy was sheer human bravery. Ambrose again captures the image:

> The beach exits from Omaha were five draws, trails cut over centuries by streams flowing into the Channel from the cliffs above. The draws were mined and well registered by enemy mortars and artillery. As small groups of men—half a dozen here, ten or so there—began to claw their way up the cliffs, they moved between the draws. Valor and willpower triumphed where planning and prophecy failed.[6]

Storming Nasiriyah: The Left Hook

In the winter of 1991, forty-seven years after Normandy, another allied military coalition prepared to launch a massive invasion against another entrenched enemy force with the aim of ejecting it from the territory it occupied. Much had changed in the structure, combat doctrine, and military technology of the U.S. armed forces in that nearly five-decade interval, and in retrospect the conduct of the ground campaign in Operation Desert Storm demonstrated that the Pentagon had made significant progress by the early 1990s in using advanced technology and a new doctrine of maneuver instead of relying on massed frontal attack to defeat a heavily armed foe.[7]

First, contemporary technology—in the form of modern transport and aerial refueling aircraft, and fast cargo ships—enabled the allied force to deploy to the Arabian Peninsula with a speed that World War II planners could not have imagined. It took more than two years for all the U.S. soldiers earmarked for Normandy to arrive in England, all but a fraction coming by ship. During the first massive buildup in Vietnam during 1964–65, it took 365 days for the United States to send 184,000 soldiers into Vietnam. The reach of Air Force planes and chartered civilian passenger jets enabled the U.S. Central Command to deploy that same number—184,000 soldiers—to Saudi Arabia in only 88 days in 1990.[8] During the entire six-month buildup for the offensive air and ground campaigns in Operation Desert Storm, the United States was able to move 1.8 million tons of cargo, including 126,400 vehicles, 1,800 Army helicopters and small planes, another 350,000 tons of ammunition, and more than 350,000 military personnel over 8,000 miles from the continental United States to the Arabian Peninsula.[9]

Unlike his famed predecessors from the World War II military high command, General H. Norman Schwarzkopf arrived in his theater of operations in the fall of 1990 relatively well informed about his enemy's location, activities, and anticipated course of operation. The U.S. Central Command from the outset enjoyed two significant advantages over the Iraqi Army: it was deployed for the most part on a flat desert terrain, so it was exceedingly difficult for the Iraqis to cloak their movements; and the Iraqi force was designed after the Soviet military model, which emphasized a doctrine of rigid centralized control from the top and little initiative by lower-level military commanders.

Schwarzkopf's goal of attaining *dominant battlespace knowledge* was aided primarily by the many tools at his disposal. To track Iraqi ground forces, Scud missile launchers, Iraqi military communications, combat aircraft, and warships, Schwarzkopf could call on a constellation of reconnaissance satellites, spy planes, ground sensor stations, and other eavesdropping facilities manned by various U.S. intelligence agencies. To deny the Iraqis that same knowledge, upon commencing its Desert Shield military buildup Central Command had immediately achieved air superiority over the Arabian Peninsula, thus preventing Iraq from conducting its own aerial reconnaissance of the allied forces. When the air campaign began on January 17, 1991, coalition aircraft equipped with antiradiation missiles and powerful jamming pods blinded the Iraqi air defense network, and strike aircraft systematically dismantled the Iraqi radar sites, missile batteries, and command centers.

For the next two weeks the allied coalition successfully shifted 255,000 soldiers in the heavy armored units of VII Corps and the "light" forces of the XVIII Airborne Corps more than 300 miles to the west—with confidence that the mass movement would go undetected by Iraqi military commanders—to position the force for the now-famous "left hook" sweep into southern Iraq.

By late February 1991 Schwarzkopf also enjoyed the means to push for *near-perfect mission assignment* using a modern, multilayered communications network in the region to provide the necessary voice, data, and imagery transmissions between his command headquarters and its various component commands. Beginning with an ad hoc satellite communications network and a small army tactical radio system, the command-and-control mechanism had grown as the troop buildup progressed to become the central nervous system for what one observer called "a machine with 300,000 moving parts."

Through secure communications circuits available to the U.S. military in the early 1990s, Schwarzkopf and his staff were in constant touch with the far-flung units, including the naval forces in the Persian Gulf and Red Sea, and the general was in daily telephone contact with General Colin Powell, chairman of the Joint Chiefs of Staff, in Washington, D.C. (and many of Schwarzkopf's subordinates were in contact with their Pentagon staff counterparts). Although not extraordinary today, this capability stands in stark contrast with the limited communications available to Allied commanders in World War II. At the time of the Operation Desert Storm ground campaign, the surviving Iraqi command centers had imposed a moratorium on electronic transmissions that was so severe (to avoid detection and air strikes) that elements of the Iraqi Army were functionally unable to respond and maneuver to counter the allied attack.

The ground offensive itself was a preordained victory given the superior training and capabilities of the American soldiers and the technological superiority of U.S. military weapons systems, particularly the sensors and night-vision devices that enabled them to "own the night."

As far as *battle assessment* capabilities, the allied coalition managed a fleet of reconnaissance and surveillance aircraft and received downloaded satellite photographs from U.S. intelligence agencies that enabled them to conduct follow-on air strikes at a pace that had never been attempted before. Both the experimental Joint Surveillance Target Attack Radar System (JSTARS, installed aboard converted Boeing 707 aircraft) and the well-established Airborne Warning and Control System (AWACS) aircraft provided the air and ground commanders with a real-time glimpse of the airspace and battlefield never before enjoyed by any military commander.

Given the combination of highly trained U.S. and allied military personnel, a 1980s-era doctrine of AirLand Battle that incorporated air and ground military power in a coherent team, and vastly superior military equipment—it was no surprise that the U.S.-led coalition was able to rout the Iraqi Army in the one-hundred-hour ground campaign.

And the starkest measure of American military progress between 1944 and 1991 came in the ultimate invoice: *the number of our soldiers and sailors killed in battle.* The allied ground forces at Normandy and Desert Storm were roughly equivalent in size, about 650,000 troops. The United States and its allies suffered over 15,000 casualties at Normandy, including 4,900 killed in action. In Operation Desert Storm there were 1,148 allied casualties, including 247 killed—compared to tens of thousands of Iraqis who were killed or wounded. Not only did modern military technology produce victory; it saved lives.[10]

THE TRANSITIONAL WAR

Looking back on the Gulf War victory in 1995, then Secretary of Defense William Perry praised the "revolutionary advance in military capability" provided by a new class of military systems that had seen their debut in the conflict with Iraq. "Key to this capability is a new generation of military support systems—intelligence sensors, defense suppression systems and precision guidance subsystems—that serve as 'force multipliers' by increasing the effectiveness of U.S. weapon systems," Perry wrote in *Foreign Affairs.* "An Army with such technology has an overwhelming advantage over an army without it, much as an army equipped with tanks could overwhelm an army with horse cavalry."[11]

As undersecretary of defense for research and engineering during the late 1970s, Perry had pushed for the development of stealth aircraft designed to counter advanced Soviet radar systems. In his 1995 article Perry succinctly noted that it was not new weapons systems alone that made the difference in Operation Desert Storm; rather, it was the array of "support systems" that provided Schwarzkopf and his lieutenants increased battlefield situation awareness and enabled them to use the new generation of precision-guided weapons to maximum effect. The "fog of war" in Operation Desert Storm was greatly reduced both by the sensors that allowed allied forces to see the enemy better, and by the intense pre-combat training that gave U.S. troops the psychological edge over their confused and isolated Iraqi opponents, Perry wrote.

But we should not construe the mismatch between Iraq and the Western coalition as evidence that the American Revolution in Military Affairs had fully arrived in 1991. Important technological elements were present on the battlefield. It was indeed the first major conflict in which stealth aircraft and precision-guided munitions played a visible and significant role. And this engagement has served as a baseline for those of us who are pressing to leverage America's information-age technological superiority into the armed forces of this new century.

To the layperson, the television images from Operation Desert Storm augured a conflict out of a science fiction novel, particularly the gun-camera images of laser-guided bombs striking targets with pinpoint accuracy, and the eerie images of combat seen through night-vision goggles. But the televised glimpses of that war were misleading in that the conflict more closely resembled traditional battlefields of the past rather than the true Revolution in Military Affairs that is now possible.

In my assessment, Operation Desert Storm was really a *transitional war* that spanned the industrial-age military tenets of mass, redundant weapons and units, and attrition on the one hand, and the information-age precepts of maneuver, communications networks, and precision-guided ordnance on the other hand. The United States had indeed deployed a massive military force halfway around the world in one-quarter the time it took to assemble the same number of personnel in England for the invasion of France in 1944. And the Central Command planners were indeed able to compress the time frame of the military campaign to a short but intense five-week air war and one-hundred-hour ground offensive. And, as I will detail below, the United States sent an entire new generation of military technology to war for the first time. But still, the Desert Storm operation relied on *mass*—the deployment of 10 U.S. Army and Marine Corps divisions, over 4,000 combat aircraft, and over 100 Navy warships—to provide the guarantee of overwhelming combat power despite the "force multiplier" effect of the command's space-age communications networks and smart weapons.

Sober analysis of the conduct of the Gulf War reveals many shortcomings in the way in which the information-age technology was employed. It is important to acknowledge these shortcomings in order to place Operation Desert Storm in its proper context as a conflict falling midway between the world of industrial warfare and the twenty-first-century information age where the Revolution in Military Affairs will take full effect. The shortcomings included[12]

- *Not enough smart weapons.* Eight types of precision-guided gravity bombs and missiles were used extensively in the 1991 conflict, including the Tomahawk cruise missile, the Paveway series of laser-guided bombs, the High-speed Anti-Radiation Missile (HARM), and the Maverick and Hellfire air-to-ground missiles. But although used with impressive results against high-value targets, precision-guided munitions constituted only 9 percent (7,400 tons) of the 84,200 tons of aircraft munitions expended. Thus 72 percent of the air war was waged as in World War II, with saturation bombing against area targets.[13]
- *Too much data, not enough analysis.* The array of national intelligence systems and tactical military sensors continuously generated great volumes of raw data about the Iraqi forces, their locations, activities and movements, and communications—almost too much data. Combat units deployed to fight the Central Command ground offensive did not have enough people or equipment to process the raw data into meaningful depictions of the Iraqi Army, its combat capability, and its current and projected course of movement. This situation further deteriorated once combat units stormed into Iraq during the ground offensive. As an official U.S. Army after-action report noted, "Dissemination proved to be the Achilles heel of the intelligence system in Desert Storm. Intelligence was generated in such great quantity that existing communications proved incapable of pushing the required hard-copy imagery showing enemy forces or emplacements, and information such as intelligence reports, communications intercepts or revived maps for a new course of advance down below the division level."[14]

- *Incompatible communications links.* The U.S. Air Force managed the air bombing campaign against Iraq, coordinating a complex matrix of targets and aircraft assignments with Air Force, U.S. Navy, and allied air units operating over 1,200 strike aircraft. The daily "air tasking order" consisted of a three-hundred-page document listing specific targets and units assigned to strike them. The sheer size of the plan created a logistical headache even for units that could receive the material via secure electronic communications networks. But for the U.S. Navy, it was worse; their systems were incompatible with the Air Force's, necessitating daily shuttle flights carrying paper copies of the tasking order from Riyadh to the six aircraft carriers in the Red Sea and Persian Gulf. As a result, the Navy flew fewer direct combat missions than it was physically capable of performing and was unable to fully coordinate its combat aircraft with those of the ground-based Air Force.
- *The targeting controversy.* Although the Persian Gulf War constituted a major improvement in interservice cooperation over previous wars, the four armed services still operated under separate combat doctrines, leading to intense and bitter conflict between the Army and Air Force over the selection of priority targets to strike. The Air Force, whose doctrine envisions using strategic and tactical airpower to attempt destruction of the enemy's total ability to wage war, insisted on attacking targets deep inside Iraq considered vital to that goal. The Army, focused on the upcoming ground offensive, wanted the Air Force to concentrate on hitting Iraqi Army units in the planned line of advance of the ground forces. Determining the targets to be struck was complicated by a shortage of aircraft dedicated to tactical reconnaissance. At one point the Army ground force operations officer sent an internal message that summarized the ground force's bitterness: "Air support-related issues continue to plague final preparations for offensive operations and raise doubts concerning our ability to effectively shape the battlefield prior to initiation of the ground campaign. . . . Army-nominated targets are not being serviced."[15]
- *Battle-damage assessment dispute.* The single-most controversial dispute among players in Operation Desert Storm centered on efforts to assess the damage inflicted by the allied air forces on Iraqi

targets. To estimate whether a target had been destroyed, the U.S. intelligence agencies, relying on satellite imagery, used a formula that was more conservative than the one used by the tactical air commanders, who relied on cockpit video recordings of bomb and missile strikes and photos taken by reconnaissance aircraft. At one point in the Desert Storm air campaign, Central Command air staff estimated that over 1,400 Iraqi tanks had been destroyed. Relying exclusively on satellite images and a more conservative equation for estimating bombing kills, the CIA reported to President Bush that it estimated that only 358 tanks had been destroyed in the same effort. The issue was far from academic, since the estimate of damage drove both the day-to-day tactics of the air forces and the planning for the time and sequence of the ground offensive. The dispute lingered long after the war itself, with Schwarzkopf bitterly assailing the Central Intelligence Agency and the National Reconnaissance Office in congressional hearings later in 1991. The disagreement stemmed in part from a failure of the intelligence agencies and the tactical military commanders to agree on the visual clues from precision-guided weapons, which frequently destroyed the interior of an underground facility without causing visible damage to the exterior of the structure.[16]

- *The escape of the Iraqi Republican Guard.* The overall goal of the ground offensive against the Iraqi Army was articulated by Joint Chiefs Chairman General Colin Powell in thirteen chilling words: "First we're going to cut it off, and then we're going to kill it."[17] But as the ground offensive entered its third and final day, a seam opened in the pincer movement between the VII Corps and XVIII Airborne Corps as they closed in on Iraqi Republican Guard units in Iraq, and about half of the Iraqi Army's surviving main force—three of six Republican Guard divisions—escaped across the Euphrates River. The matter remains one of the most controversial issues of Desert Storm and has been debated in the various official and outside accounts of the conflict. The most credible explanation comes in *The Generals' War*, the account written by *New York Times* national security correspondent Michael R. Gordon and retired Marine Lieutenant General Bernard E. Trainor, who attribute the incident to two factors: The Marine Corps's direct offensive into

> southern Kuwait succeeded more quickly than Central Command planners had anticipated, and the vast "left hook" pincer movement by the two Army corps could not close the gap in time even though their attack had been moved up by about fifteen hours from the original plan. Also a combination of bad weather and obscuring smoke clouds from burning Kuwaiti oil fields prevented the coalition reconnaissance efforts from detecting the Iraqi retreat in time to block it. This incident above all others underscored the continued presence of the "fog and friction" of war, as elucidated by von Clausewitz nearly two centuries earlier, on the Desert Storm battlefield, despite the stunning array of high technology designed to remove that fog and friction.

None of this is to ignore the fact that many new combat systems performed admirably in Operation Desert Storm. The F-117A stealth fighter served with distinction in pinpoint bombing raids over heavily defended targets. The Tomahawk cruise missile was effective and accurate at long range. JSTARS, although still in development and installed onboard only two aircraft, provided a unique real-time view of enemy units on the move despite the thick clouds that cloaked the battlefield. Both modern technology and traditional American innovation mastered the innumerable complexities involved in moving a force the size of the population of Atlanta halfway around the world in six months—and then getting them home again once the war was over. But the Persian Gulf War was not a validation of the Revolution in Military Affairs; it was, instead, merely a harbinger of the technological promise now before us and the extraordinary increase in military power and intelligence that the revolution will provide us if we can successfully carry it out.

ELEMENTS OF THE REVOLUTION IN MILITARY AFFAIRS

Battle awareness—knowing the location of both friendly and enemy units, weapons, and soldiers—has always been a critical element of military operations. But new technologies can help transform that awareness to *dominant battlespace knowledge*; that is, technologies can raise an awareness of where things are to an understanding of the relationships and relative significance among them—how units and weapons on the battlefield fit into and are constrained by the terrain, what they are trying to accomplish, and how they relate to each other. This kind of knowledge constitutes an insight into the future, for it enables us to understand how the enemy commander sees his own battlefield options, and therefore increases the accuracy of predicting what he will try to do. Although great military leaders have predicted enemy movements and tactics intuitively, the ability to superimpose the location of various units over the terrain in which they are operating increases the accuracy of such predictions. Combining that ability with an understanding of the command relationships among the units adds more predictive accuracy. The array of communications and information-gathering technology available now or in the near future—imagery from satellites and unmanned aerial vehicles, advanced electronic signal intelligence gathering, and the computer networks and communications capability to process and disseminate the information throughout the U.S. force—will provide the U.S. commander with this knowledge.

This step from battlefield awareness to knowledge will allow the U.S. military commander to apply the right kind of force at the right time against the right targets. It provides a basis for allocating precision-guided munitions rationally, in ways that can increase their effectiveness. It is the foundation of effective targeting not just against fixed sites, but directly against an opponent's military units. And it is the best way of avoiding fratricide on the battlefield, those most horrible incidents when we accidentally kill our own troops.

Finally, the "system of systems" at the heart of the Revolution in Military Affairs promises a new capacity for the U.S. commander—knowing the results of his military operations almost immediately, which is central to building a military that can adapt to a highly complex situation faster and better than an opponent can. With this information the U.S. forces can operate within an opponent's decision and action cycles—outthinking and outmaneuvering the enemy commander. If we are able to make the right choices faster than an opponent can and we can move more agilely, we will be able to win in any kind of military confrontation.

Part 1: Seeing the Battlespace

Intelligence collection, surveillance, and reconnaissance, known by its military acronym ISR, is a key component of the "system of systems" that provides the means for the U.S. military commander to see the entirety of the battlespace. Current and future ISR capabilities stem from two long-term trends in military technology. First there is the growth of various collection "platforms," such as orbiting space satellites and specialized surveillance aircraft, submarines, and ground sensors. The second trend is the continuous improvement to that array of platforms by new generations of sensors that span an increasingly broad width of the electromagnetic spectrum. The "revisit" times by which we can look, listen, measure, take the temperature, register the magnetic properties of the battlespace, and test for changes have diminished, and the *take*—the resulting intelligence—gets more and more comprehensive.

For purpose of discussion, picture an area of 40,000 square miles—an area roughly the size of the Kuwaiti theater of operations during Desert Storm, or the land and sea area between Seoul, South Korea, and P'yŏngyang, North Korea. This fictional terrain is a composite of sea littoral, desert, forests, and mountains, and it will serve as a reference as we describe what we will be able to see and understand about a battlespace over the next decade if we improve our surveillance capability in the ways I am calling for here.

The technology available today and that which will enter the inventories in the next several years will make a revolutionary difference in our capacity to understand what takes place in such an area. By 2010—and earlier if we accelerate the current rate of research and procurement—the U.S. military will be able to "see" virtually everything of military significance in and above such an area all the time, in all weather conditions, and regardless of the terrain. We will be able to identify and track—in near real time—all major items of military equipment, from trucks and other vehicles on the ground to ships and aircraft. More important, the U.S. military commander will understand what he sees.

Dominant battlespace knowledge begins with situation awareness, the commander's ability to see things of military significance—everything from tanks and missile bases to terrain features such as cliffs and other natural features that might shelter enemy forces or weapons. All militaries have this ability to a significant degree, but there are two characteristics about the U.S. military today that stand out. The first is the marked and growing disparity between the ability of the U.S. military and other nations' forces, largely because of our technological superiority and our greater investment in air- and space-based observation systems. The second is the military capability this technology provides our commanders.

Space-based Observation

One specialized area in which the U.S. military—led by the Air Force—is quietly advancing the current military revolution is in space. This revolution has been ongoing since Project Bumper, the first successful launch of a two-stage rocket 244 miles up into space on February 24, 1949, and has spanned the era of manned spaceflight. But until very recently, the U.S. military's accomplishments in space were cloaked by the highest levels of official secrecy.

The military role in space today, as it was also four decades ago, is focused on the use of unmanned satellites to conduct reconnaissance on military sites below. Of 1,679 military and civilian payloads sent into orbit over that forty-year period, 431 of them have been satellites employed for surveillance and intelligence gathering.

Space remains the key strategic environment for winning the current Revolution in Military Affairs against any enemy. The long U.S. technological superiority in space was a major contributing factor in our bloodless Cold War victory against the Soviet Union, since our constellations of photo-reconnaissance, surveillance, and electronic spy satellites provided U.S. leaders an adequate knowledge of Soviet military activities to avoid inadvertent war. (The best example of this capability came during the Cuban Missile Crisis of 1962, when our space-based surveillance systems assured then President John F. Kennedy that the Soviet Union—despite the bombastic rhetoric coming out of Moscow—had only a handful of intercontinental ballistic missiles capable of striking the U.S. mainland, and even these had not been put on any combat alert. This knowledge freed Kennedy to veto a recommended U.S. military invasion of Cuba and instead opt for a naval "quarantine" and subsequent negotiations with the Soviets that ended the confrontation without a military clash.)

Space surveillance and communications satellites now and in the future will play an even more central role in U.S. military operations.

Space satellite sensors are not bound by rules of national sovereignty. Unlike sensors carried by aircraft or used on the earth's surface, these devices ignore political boundaries and need no "clearance" from nations whose territory they wish to see.

Most U.S. surveillance and reconnaissance satellites fly either in low-earth orbits between 100 and 300 miles above the earth, or in stationary geosynchronous orbits at an altitude above 20,000 miles, which keeps the satellites positioned continuously over a fixed point on the earth's surface. There are advantages and drawbacks in the current technology employed in each of the types of satellites used to spy on the earth's surface.

Low-level satellites can take detailed pictures in the visible, infrared, and microwave bands and employ either optical cameras or onboard radar transmitters to obtain their images. But one disadvantage is that coverage from low altitude cannot be constant owing to the unalterable laws of physics and gravity that predetermine a satellite's orbital parameters. Although these reconnaissance and surveillance platforms can view a swath of the earth's surface in great detail as they revolve around the planet, there are inevitably gaps in coverage that a smart enemy can employ to his advantage.[18] (Some sensors on U.S. low-earth orbit satellites do have the ability to "stare" at a particular area for as long as the area is in view from the vehicle carrying the sensor, as well as adjust their scan to the left or right of the orbit path, providing some flexibility in surveillance.)

Satellites in geosynchronous orbit, on the other hand, observe large areas of the earth's surface continuously. But to maintain such orbits they must operate at considerable distances from the surface—about 22,300 miles above the earth's equator. Generally, sensors in geosynchronous orbits collect electronic and infrared emissions, tracking, for instance, the heat plumes of boosting ballistic missiles or the thermal pulse from nuclear detonations. Weather satellites that provide the daily images seen on most local television news reports use electro-optical devices from their positions in high geosynchronous orbit.

Our space-based surveillance and reconnaissance capabilities have expanded continually since the first primitive optical sensors went into orbit nearly forty years ago. The U.S. space-based defense support system was initially limited to optical sensors collecting information transmitted by visible light. Development of these early satellites accelerated after the May 1, 1960, shootdown of the CIA-operated U-2 reconnaissance aircraft piloted by Francis Gary Powers. The optical systems they carried were remarkable for their time.

Yet, like airborne cameras, the optical systems could usually collect information only during daylight hours and in clear weather. Infrared collectors enhanced our ability to collect images by providing satellites with the means to identify objects by their heat signature. This technology also had its disadvantages, since infrared signals from the ground were still obscured by weather conditions such as clouds, rain, and snow. Space-based radar sensors, picking up the echoes of the radar pulses transmitted to the ground from space, finally overcame the obstacles posed by weather conditions and—to an extent—enabled us to "see" through foliage and other cover.[19] And because satellites used to collect radio signals and other electronic emissions are of high value to U.S. military and intelligence agencies, the effort to use satellites for these purposes has expanded continuously and significantly over the past three decades.

Meanwhile, the United States has been deploying constellations of space-based sensors, coordinating their orbits in ways that have provided wider area surveillance of strategically important regions such as the Soviet Union, China, and North Korea. But there are still important gaps in our ability to perform reconnaissance from satellites in space. It is still difficult to capture radio beams at particular frequencies, including those used by the increasingly ubiquitous cellular telephone, and other parts of the electromagnetic spectrum remain completely invisible to U.S. surveillance satellites. It is not yet possible to collect *all* the electromagnetic emissions *continuously* from any given area 200 miles by 200 miles on the earth's surface. This means that some areas of the world remain relatively uncovered and that no area is covered continuously.

This limitation on coverage is largely because of the tremendous costs of the satellites themselves and the effort to lift them into space aboard either the space shuttle or unmanned rockets. The Pentagon's current top-of-the-line Milstar communications satellite costs about $800 million, and a classified surveillance satellite lost in a 1998 rocket explosion was reported to cost at least the same.[20] Nor is it cheap to build and lift sensors into space. It costs about $10,000 per pound to put payloads in orbit using either the shuttle or planned unmanned satellite launchers. Building the ground receivers and infrastructure needed to make use of the information collected by satellites does not come inexpensively either. What we have spent to achieve our current capability remains a classified figure, but it is certainly in the multiple billions of dollars.

Thus over the years the United States has had to impose some tough priorities on its space-based surveillance programs, focusing on what we identified as the strategic threats to the United States first, and later developing the capability to use the entire range of our space-based information collection assets for tactical military purposes.

As our capability to collect information has expanded across the electromagnetic spectrum, we have focused on building an integrated system by fairly systematically tying the different sensors together. For example, most of the optical sensors are designed to provide relatively high resolution of what they "see," although the high resolution severely restricts their field of view, as is the case with a large telephoto lens. Although it is conceivably possible—but ruinously expensive—to photograph every square foot of a wide area of the earth's surface with enough passes of a satellite, a more cost-effective and less time-consuming approach is to use sensors with a wider field of view to scan a large area and upon detection of a possibly suspicious target, to bring in the more powerful sensor with the narrower optical perspective.

But even now, we still have serious gaps in our overall ability to conduct detailed surveillance of the earth's surface. Figure 3-2 depicts my latest estimate of our various satellite sensing capabilities for a hypothetical 200-by-200-mile area of the earth's surface, representing the

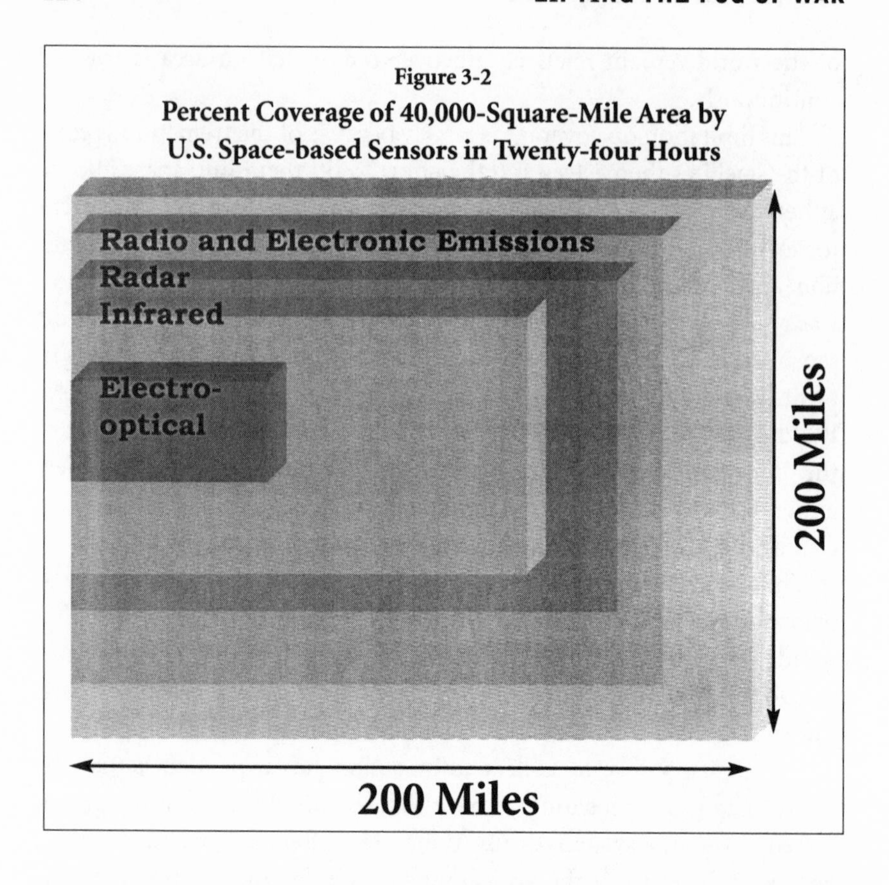

Figure 3-2
Percent Coverage of 40,000-Square-Mile Area by U.S. Space-based Sensors in Twenty-four Hours

notional crisis region or extended battlefield of the near future. That area is represented by the outer square. The smaller rectangles roughly indicate the portions of that same geographical area that various space-based U.S. sensors could observe over a period of twenty-four hours.

The areas depicted on this figure are optimistic because, among other things, the surveillance coverage depicted here is calculated based on the assumption that the earth's surface is smooth and uncovered by vegetation and man-made structures. In reality, areas the U.S. needs to monitor are replete with hills, mountains, valleys, vegetation, and structures that shadow or cloak what lies beneath them from the view of many of our orbiting sensors. So what space-based sensors see depends a lot on the particular area that is under surveillance. More can be seen in a desert than in a mountainous jungle. More can be seen on the ocean's surface than in an urban area.

But even with these limitations, it's hard for our contemporary foes to avoid U.S. surveillance altogether. If a large-scale military operation does try to avoid detection, in doing so it renders itself less effective in other ways. The measures an enemy takes to avoid being seen—from hiding in the "shadows," to reducing electronic and radio emissions, to deploying decoys—can slow and disrupt movements of combat units and materiel and make it much more difficult for the forces to conduct coordinated military operations. And it will be increasingly difficult to hide from U.S. space-based observation in the future because the United States is continuously working to improve the quality and coverage of its space-based surveillance.

Overcoming some obstacles is simply a matter of making informed use of the raw data. For example, we can take advantage of some of the same phenomena that limit sensor coverage. Hills and valleys create shadowy pockets where various sensors cannot see. But these topographical features can also create areas where some kinds of weapons, units, or military activities are unlikely to be. It would be unusual to find armored vehicle assembly areas in the middle of a lake, for instance. Today, as ever in the history of warfare, "taking" a territory depends in great part on understanding its topography and putting this knowledge to one's own use, rather than the enemy's. And our understanding is growing, in large part because of the success the United States has had in mapping the surface of the earth from space.

The engine of this success is the capacity to measure precisely with radar the elevation of points on the earth's surface. The United States has been doing this through the use of digital topographical mapping for several years and is now beginning to do so with a resolution level accurate to 5 meters or less. Working from the elevations of several known points—say, the peaks of two mountains—programs can interpolate the elevation of all intermediate points, until the computer generates a line indicating the lay of the land, which in turn enables the programs to assess the actual gradients and slopes of the terrain. This, in turn, allows us to exclude some of the areas that are obscured from sensors because they are unlikely to harbor targets of military significance.

We can also focus our surveillance effort from space by cross-cueing the various sensors we have. Sensors in geosynchronous orbits are able to cover large areas continuously. They may not be able to identify the source of the electromagnetic signals they collect and they may not be able to pinpoint precisely the transmitter location, but they can come close enough to permit more accurate, long-range sensors to actually find the target. This is a way of strengthening the capability of space-based coverage without having to launch a massive and prohibitively expensive fleet of additional satellites.

As the Revolution in Military Affairs matures, we are poised to deploy new classes of space-based sensors according to the high priority the Department of Defense has assigned to U.S. space operations. In his *Annual Report to the President and Congress* two years ago, Secretary of Defense William Cohen explained the Pentagon's continued rationale for this capability and the need to modernize our abilities with the planned "Space-Based Infrared System" satellite constellation to meet the Pentagon's surveillance requirements in the new century:

> Space forces provide the sole means to access otherwise denied areas of foreign countries without violating their sovereignty. The command, control, communications, computers, intelligence, surveillance, and reconnaissance (C^4ISR) capabilities provided by space forces are crucial to generating information necessary to support investment decisions that maintain U.S. military preparedness and readiness, to support military planning, and to enable information superiority during a crisis or conflict.[21]

The Pentagon has developed a new satellite system that will use infrared detection sensors on orbiting satellites in a variety of locations. The Space-Based Infrared System (SBIRS) program will constitute a single "system of systems" supporting missile warning, missile defense, and intelligence applications. The first launch of a geosynchronous (SBIRS-High) satellite is scheduled for 2002. The second component (SBIRS-Low) will provide unique midcourse tracking of threats, a capability that will significantly enhance performance of both theater and possible national missile defenses. The SBIRS-Low concept calls for a constellation of twenty-four satellites working synergistically with the SBIRS-High satellite. The first launch of the SBIRS-Low is scheduled for late 2004.

The emergence of space as both a military environment and a vital commercial arena is central to the Revolution in Military Affairs—and space capabilities are going to happen whether or not the United States and its allies decide to seize the opportunities. The Pentagon has already girded itself to regard space as an area that must be guarded and defended. In his annual report in 1998 Secretary Cohen warned,

> While U.S. national security interests focused in the past on assuring the availability of oil, the future may require greater interest in protecting and accessing the flow of information. As a result, the importance of space as a principal avenue for the unimpeded flow of information throughout a global market increases. DoD recognizes these strategic imperatives and will assure free access to and use of space to support U.S. national security and economic interests.[22]

As governments and corporations begin to take greater advantage of space, so-called "civilian space imagery" will become more and more abundant. Like other countries, the United States has several new commercial organizations that are using private satellites to create their own photographic and radar images from space for commercial and private use. Since 1991 a vigorous market demand has developed for satellite imagery from Russian satellites that have an accuracy of 24 meters' resolution. The market will probably expand in the future as several private U.S. satellite imaging firms, as well as Japanese corporations, and, perhaps, a Franco-German consortium, offer space-based imagery for sale. The Japanese offer imagery resolved down to 3 meters using synthetic aperture radar, and an Indian satellite already in operation can resolve down to 6 meters using panchromatic sensors. France is working with Germany to launch the radar sensor Horus satellite and the Helios 2 satellite with electro-optical imaging down to 1 meter.[23] Commercial applications include land-use management, environmental monitoring of areas vulnerable to pollution, commercial photography, and scientific research into the impact of long-range weather patterns.

There will continue to be some significant limitations to space-based surveillance in terms of how much information it provides and how precise, accurate, and timely that information is. But space-based observation has remarkable advantages, too: the higher the vista point, the greater the distance that can be seen, the broader the view, and the larger the number of things that are likely to be seen. This tool has obvious military importance, particularly if the sensor data can be converted quickly to an information format such as a new map of the battlefield or a refined estimate of enemy troop strength that provides a military commander greater understanding of events in the area under surveillance. Given the inherent connection between spatial distance and time in military affairs, the view from space will provide the U.S. commander a genuine view into the future. The capacity to see moving enemy units before they come close enough to threaten U.S. forces constitutes a valuable glimpse into the future. Forewarned is forearmed, as the saying goes, and satellite surveillance promises the kind of forewarning the great military tacticians of the past could only dream of.

But space-based surveillance is not by itself a sufficient foundation on which to build dominant battlespace knowledge, let alone to protect American interests or to win wars. For although we may be able to see an object at a great distance, we may not, because of the distance, be able to tell what it is. What looks like a tank convoy may be a row of rusted hulks of old transport carriers; what looks like a munitions factory may be a dairy processing plant, and so on.

Airborne Observation

The use of manned aircraft for reconnaissance is a practice nearly as old as the manned aircraft itself. Only eleven years after the Wright Brothers got the first airplane aloft at Kitty Hawk in 1903, both sides in World War I were using biplanes to conduct aerial surveillance of enemy trenches and troop movements. The U.S. Air Force began as the Aeronautical Division of the U.S. Army Signal Corps, whose primary mission was to survey the battlefield and offer communications support to the forces on the ground, performing battlefield observation and serving as communications support for the ground forces.

Nearly a century later, airborne reconnaissance and surveillance still play a central role in military intelligence operations. And even as our armed forces have made a concerted push into space over the last forty years, we have improved our flight intelligence exponentially. The two most prominent examples are the E-3 AWACS aircraft (which can monitor the air battlefield for thousands of square miles and manage hundreds of fighters as they patrol and fight) and the JSTARS ground radar surveillance aircraft (which has been used in both Operation Desert Storm and Kosovo to locate and identify enemy troop movements on the ground).

The advantages and limiting factors of aircraft surveillance are different from those of space-based platforms. Unlike satellites in space, aircraft have to pay attention to the sovereignty of national borders. This does not mean that political boundaries necessarily block what aircraft sensors can see, or that a nation that denies access to the airspace over its territory is immune from airborne observation and surveillance. Some airborne sensors can look across borders deep into the airspace and territory of another country.[24] Yet in peacetime there are going to be limits to the extent to which airborne sensors can provide broad area coverage where there are political problems with crossing borders.

Conflict is another matter. During a conflict, the political or legal constraints on flight access no longer apply, but the risk is naturally greater. Concerns for the aircraft can be valid and important, for many of the most capable airborne sensors in the U.S. inventory are carried by aircraft that are not very stealthy and that must emit energy to generate the echoes they read. As more countries acquire the sensors to find and track these platforms, as well as missiles of sufficient range to attack them, the use of manned surveillance aircraft will become more complicated and the need for supplementary defensive resources will increase.

In response to aircraft vulnerability, the Defense Department in recent years has given priority to developing new generations of Unmanned Aerial Vehicles (UAVs). These airborne vehicles have been around since the 1960s, but within the last five years several technological advances—in propulsion, sensors, guidance, and in the vehicles' capacity to transmit images—have driven them to the forefront of the U.S. surveillance effort. Originally viewed as "special mission" platforms to be sent where the risk of sending manned aircraft was too high, they carried optical sensors that recorded data to be processed once the vehicle returned to its launch point or landed elsewhere. Processing the film or videotapes carried by these Vietnam-era surveillance and reconnaissance UAVs normally entailed time delays that diluted the value of the information they had gathered, so the early model reconnaissance drones were usually employed on missions to collect information to be included in strategic databases of fixed target sites.

The advent of new micro-processing units, small and powerful enough to allow onboard image processing and transmission, transformed the UAV from a minimally useful piece of equipment to a new form of surveillance technology that promises to be a major "force multiplier" in future operations. Current UAVs carry off-the-shelf video cameras that can track targets below and sophisticated radio transmitters that can send the data directly to ground controlling stations. Today we already use UAVs as an integral component of any U.S. surveillance network, where their ability to provide information about suspicious activities initially detected by other sensors is invaluable. Their advantages over manned aircraft are obvious: in addition to being less expensive, they can stay aloft longer without refueling, can loiter over targets, and can operate under cloud cover that would thwart manned reconnaissance aircraft. UAVs can provide high-resolution images of relatively small areas of interest. The same relatively light sensor package that yields 10-meter resolution from a reconnaissance satellite can, if installed on a UAV, yield a photograph whose resolution of 0.1 meter from a few miles above the earth's surface is a hundred times as sharp as that of the space sensor image.

Currently there are three classes of American UAVs: tactical, mid-altitude (and endurance, for extended flight times), and high-altitude (and endurance). The United States has concentrated UAV development largely on fixed-wing vehicles—on planes rather than helicopters. This choice made sense in the past because fixed-wing aircraft have generally been able to fly farther and faster than either rotary-wing vehicles or lighter-than-air platforms such as blimps, a factor that offered the advantage that the fixed-wing aircraft could be based (or launched) at a greater distance from the surveillance target. Likewise, some of the sensors we want the vehicles to carry, such as synthetic aperture and moving target indicator radar, are large and heavy, and fixed-wing aircraft can usually carry larger payloads than helicopters for the same amount of fuel. Usually planes can also fly higher, something that is of value both for the wider scope of observation they can provide and for the greater survivability higher operating altitudes offer.

In the future, technological advances will enable us to supplement our growing fixed-wing UAV inventory (at the present, numbering in the dozens) with rotary-wing unmanned aircraft. Rotary-wing aircraft can hover and fly vertically, capabilities that could be of considerable help in urban areas or other challenging terrain. Technical advances in UAV components including focal-plane array technology, mechanical coolers, electronic packaging, and inertial sensor miniaturization will improve performance and lower the weight of sensors. These innovations will benefit both fixed- and rotary-wing aircraft, but they may push the cost-effectiveness of the latter to the point that they become competitors for some of the missions monopolized by fixed-wing UAVs. Small UAVs, with wingspans as small as 24 inches, could be built using today's technology.

Lighter-than-air platforms such as tethered aerostats (unmanned rigid balloons) obviously can't match the speed of winged UAVs, but they can offer longer flight times, and their payloads can be both large and heavy. These features have made them productive in stationary surveillance missions; for instance, radar-carrying aerostats were used for defensive surveillance along the Kuwaiti-Iraqi border. The U.S. Navy has also considered using aerostats carrying air-defense warning radar tethered to moving warships that provide the ships a much improved radar picture because the radar is positioned at a higher elevation than if the radar were mounted on a ship mast.

The larger UAVs can carry a wide variety of payloads. For other than close-range applications, synthetic aperture radar is the best system to use. This is an advanced form of radar that processes the reflected signals electronically in such a way as to replicate a much larger receiving antenna than is actually present on the vehicle, enabling the system to achieve a much finer signal image that can track and identify much smaller objects than otherwise would be possible. A second form of radar that is valuable for surveillance purposes is known as "moving target indicator" radar; it can see through weather. Optical imaging provides critical information when there are no obscurations, and it is particularly adept at tracking small, covert vehicles that operate below cloud cover. UAVs can also carry detectors for nuclear, biological, and chemical agent materials, and these unmanned vehicles can deliver ground-based acoustic, seismic, and magnetic sensors to a target area.

And equipped with Global Positioning System navigational receivers and laser designators, UAVs can record and transmit data about any point on the ground with high accuracy. Using the longer-range UAVs to deliver weapons could in effect give them the attributes of turbojet-powered cruise missiles even though long-range cruise missiles were banned under the terms of the Treaty on Intermediate-range Nuclear Forces (the INF Treaty).[25] But as target-designating vehicles, UAVs could become an integral part of long-range precision-point guided systems launched from the ground or other aircraft. When the United States and NATO conducted a massive air bombing campaign over Kosovo last year, military commanders used a small UAV-force for surveillance, but (for reasons I discuss in Chapter 5) they failed to exploit this military capability to the fullest.

Ground- and Sea-based Sensors

Improved ground-based sensors provide a third general way of "seeing" the battlefield. One class of ground-based sensors in use today is part of the Nuclear Detonation Detection System, which employs satellites and sea-based hydro-acoustic sensors in order to identify and triangulate the location of any nuclear detonation. Other sensors designed for chemical and biological warfare defense are designed to detect the presence of agents in the air. As far back as Vietnam, the Defense Department employed sensors dropped in the jungle that recorded the heat and odors of passing enemy soldiers. These "people sniffers" were inadequate for the task of tracking North Vietnamese troop movements, but their technological potential was apparent even then. Densely placed acoustic microphones can find aircraft or cruise missiles that fly below radar sensors by detecting their distinctive trail of noise. Magnetic sensors can detect moving vehicles; gravimetric sensors can differentiate among empty, lightly loaded, or densely packed trucks.

Shipboard sensors operating from international waters can pick up electronic signatures, listen to port operations, oversee the flight operations of coastal cities, and, from some locations, pick up radar signatures that hug the earth. The newest category of shipboard sensors are designed to detect the presence of concealed nuclear, chemical, or biological agents by sensing the trace emissions of those materials. An obvious limitation on the effectiveness of ship-mounted sensors is that in order for the sensors to be effective for targeting any sites on land, the ship has to move to within 15 miles of the coastline. Submarines can move in toward shore in a considerably less visible manner and can plant a range of electronic sensors. A series of buoys, in sufficient combination, might create a floating radar dish strong enough to simulate today's land-based over-the-horizon backscatter radar, thereby enabling the United States to establish radar receivers virtually anywhere on the oceans, capable of receiving radar echoes generated beyond the horizon.

Sensing man-made objects on or beneath the seas from ships and submarines requires acoustic sensors. The nature of sound propagation in water, particularly influenced by factors such as water temperature, salinity, and sound "layers," often means that sensors can receive and track signals from great distances.[26] Acoustic information, like much of the data generated from the electromagnetic spectrum, can be processed to reveal important details. Engine vibrations or the acoustic signal from a submarine's propeller can pinpoint not only the particular type of ship or submarine, but the particular vessel. As in the case of radar, acoustic signals can reveal a lot of information about the target—whether or not it is moving, what direction it is moving, its speed, and any variations in its movement.

Integrated Sight

The performance of individual sensors and the specific technology on which they are built is impressive but it is the ability to achieve *integrated sight*—the stage where the raw data gathered from a network of sensors of different types is successfully melded into information—that holds the strongest potential for U.S. military superiority in the new information age. This potential capability remains the core of the ongoing Revolution in Military Affairs and will define the future military superiority of the United States in developing dominant battlespace knowledge.

Other nations have surveillance satellites and aircraft. Others can fit these platforms with "eyes" and "ears" that survey the electromagnetic and acoustical spectra. But only the United States has the ability in the near term to build a global, integrated network of sensors and communications that will enable it to generate accurate, precise, and timely battlefield knowledge anywhere in the world.

How would integrated sight occur? Consider a network of linked U.S. military sensors tracking data from a crisis area in which the Pentagon is considering military involvement. An airborne sensor using visible light discovers a moving object that an analyst, viewing an image of the object on a video screen, assigns a 50 percent probability of being an enemy tank ("It looks like a tank, but it could just as well be a truck"). Other sensors are then used to observe the moving object through the infrared and ultraviolet areas of the electromagnetic spectrum. Analysts examining that data also conclude that the object "looks" like a tank—because a tank gives off different UV and infrared signals than a half-track or truck. Then additional sensors using active radar and passive electronic signals interception also generate data that conforms to the analysts' knowledge of what a tank looks like on radar, and what electronic signals it is known to transmit.

Thus the U.S. sensor network's capacity to correlate different sensor data can confirm the true identity of the object, even though no single sensor, viewing the object at great distance, can by itself provide conclusive verification that the moving object is a tank.

Being able to see clearly at great distances in turn will improve the U.S. military's ability to overcome two other previous limitations: the difficulty of understanding the significance of complex activity (such as the difference between a convoy of refugees and an enemy army formation) over a vast geographical area; and the challenge of discerning from among the great numbers of objects and events those that are the most important to the immediate battle situation.

Overcoming these limitations will be possible only if the profusion of raw data can be quickly and effectively collated, sorted, processed, correlated, screened, and made available to the U.S. commander quickly enough for him to use it appropriately.

Integrated sight requires more than a large number of sensors, even if the range of information they collect covers a lot of different phenomena. Disparate, high-volume streams of data can sometimes be worse than nothing at all, because too much information can lead to system overload and paralysis. Data collection is only a step—an important one, to be sure—but only one part of the military commander's effort to understand events within the extended battlespace.

One of the greatest opportunities of the current Revolution in Military Affairs is to design and build computer networks that can enhance—and replace—humans in many aspects of the data "fusion" process. A particular challenge is to deploy such a network onto the battlefield by embedding it throughout the combat force that will actually wage the war, reducing the inefficiency of having the network located at the Pentagon or at far distant headquarters where it cannot help the soldiers on the ground. One of the more advanced and creative proposals for the future U.S. military use of computing power comes from Colonel Douglas Macgregor, who has endorsed the concept of a "Joint C^4I (for command, control, communications, computers, and intelligence) Battalion" in the Army of the twenty-first century that would fully integrate the information revolution into combat forces by providing the ground force commander full "connectivity" into the sensors and information networks of all of the armed forces.

During my tenure as vice chairman of the Joint Chiefs of Staff during 1994–96, research had been well under way in this area, and I found the Joint C^4I proposal more fascinating than many other elements of the emerging military revolution.

At the center of this development is what is called automatic target recognition, assigned the military acronym ATR.

Automatic target recognition is not restricted to military applications. Some of the greatest advances in using computer processing power to glean important "targets" out of vast quantities of background material have occurred in the fields of medicine and petroleum exploration. The computer software used to conduct automated mammography screening for women already surpasses physicians' ability to locate and identify nascent tumors. Unlike humans, machines don't get tired and their attention doesn't wander when they're engaged in repetitive actions. Likewise, advanced computers are faster and more effective than humans in screening seismic data to search for potential oil-bearing geological formations.

The reason automatic target recognition has come so far has a lot to do with computer processors and the efficient way in which sight and sound sensor data can be translated into bits and bytes. Machines read and interpret data better and much faster than humans can. And they can manipulate it mathematically with an unmatchable speed and dexterity.

There are still serious limits to how well ATR works. Automatic target recognition performance depends greatly on what the sensors collect. The better the resolution—in data resolution is measured in pixels per target—the better ATR routines will work.[27]

The larger the number of angles from which a sensor can collect information, the more accurate ATR will be in picking out actual targets from their backgrounds. These "angles" can be geometric. By looking at a parked aircraft from different angles, a computer can see the aircraft in three dimensions, which in turn allows the program to make more precise comparisons of the object. For example, images generated from different frequencies provide different electromagnetic "angles." We can get these electromagnetic angles from either the same sensor, such as via multiband synthetic aperture radar, or by correlating the take of different types of sensors looking at the same object.[28]

As the U.S. military transforms itself into a true force driven by computer power, our soldiers will be able to fuse target recognition evidence from multiple sensors on different platforms—viewing an area from different angles to provide in essence a continuous three-dimensional view of the battlespace and the enemy upon it—and this will be the most powerful lever of all to ensure our combat superiority.

Earlier I used the notional example of a crisis area or extended battlespace of 200 miles by 200 miles, or 40,000 square miles, as the size of terrain and corresponding airspace that the U.S. military should be prepared to master for future conflicts. How close are we to being able to see everything of military significance in that area of 40,000 square miles? I believe we are not far from achieving that capability and will close on it rapidly—if we continue to press for technological innovation and adaptation of computer power to the military force.

From Awareness to Knowledge to Dominance

If a military commander with the help of sensor networks is able to see everything of military significance in the combat zone, then he has *battlespace awareness.* But awareness—knowing where everything is—is not knowledge. *Dominant battlespace knowledge* comes from the commander's ability to merge his awareness of a battle situation based on basic information (the location of enemy units and their patterns of movement) with additional data that gives him a deeper comprehension of the enemy's intentions, planned actions, capabilities, and limitations.

There are two ways in which a commander can defeat another military force—either another nation's regular military or an irregular guerrilla force like those we faced in Vietnam and Somalia. One way is through *attrition*—warfare based on what the textbooks call concentrated firepower, massed troop formations, sequential staging of maneuvers, and overkill. Attrition aims at the steady application of military violence until the adversary ceases to possess a combat-effective military force. The other means to battlefield victory is through *maneuver*—the directing of firepower at carefully identified and selected parts of the enemy force in an effort to destroy the enemy's command-and-control structure, or disrupt the enemy's planned sequence of operations. In other words, the armed forces find the weakest part of the enemy's defense—its reliance on a single command center or its attempt to mass huge numbers of soldiers to make up for the army's lack of weaponry—and exploit it. In maneuver warfare the assumption is that there are significant differences among military targets, and that some have more value than others because their destruction will have a more severe impact on the enemy's ability to function as a viable military organization.

With dominant battlespace knowledge the U.S. military commander will be able to discern key relationships in the enemy organization, such as the critical components of the command structure whose destruction will paralyze the enemy force, or a pivotal event in the enemy's overall military campaign whose disruption will upset the enemy's strategy or battle plan. With such knowledge the U.S. military commander can greatly enhance the power and impact of his precise, long-range, highly accurate weapons. The initial air strikes in Operation Desert Storm targeted key command units of the Iraqi air defense command on the assumption that destroying these important facilities would significantly blind the hundreds of enemy missile batteries and antiaircraft gunners, rendering them ineffective against the oncoming wave of U.S. and allied strike aircraft.

The ability to achieve dominant battlespace knowledge gives the United States and its military commanders the unprecedented opportunity to abandon the inefficient, costly, and bloody strategy of attrition warfare that has dominated the U.S. military for a century and a half from Cemetery Ridge at Gettysburg to Mutlaa Ridge in Kuwait. Replacing that obsolete and unwieldly paradigm of combat with the promise of true maneuver warfare driven by computer processing power is within our reach today.

Part 2: Communicating Dominant Battlespace Knowledge

Knowledge is power in military operations only if it can be communicated to combat forces that can use it, a capacity that depends on the effectiveness of the network of communications systems that constitute the second component of the "system of systems" underlying the American Revolution in Military Affairs.

The systems that make up the network of communications tie sensors to humans, sensors to other sensors, and humans to weapons. And because for the United States the focus of military operations is probably going to be overseas, our armed forces will continue to have some unique communications needs. Since U.S. forces are expeditionary in character, they not only have to be interconnected with each other; they must also rely on timely access to databases around the world. Further they need massive communications channels, usually characterized by large bandwidths, or system capacity, to handle certain categories of information such as fine-grain resolution pictures or other graphic material, video teleconferences, and a torrent of data.

The U.S. military is on its way to developing a true worldwide communications network to handle a wide range of command and control functions for the armed services. The Global Command and Control System (GCCS), which replaced the 1960s-era Worldwide Military Command and Control System, is operational at nearly seven hundred U.S. military sites in the continental United States and overseas. The system embraces a wide variety of communications devices and formats, including secure teletypewriters, facsimile transmissions, graphics and video channels, and voice radio networks. As configured, the new command-and-control system provides regional commanders information on force status, intelligence support, enemy order of battle, related facility information, and air tasking orders. In 1998 an updated version of the system went on line providing transmission of imagery and meteorological and oceanographic data. Part of the network is configured to handle top secret material for command and control of U.S. forces being deployed overseas. The Defense Department has committed itself to developing these "battle management systems" not only for the Pentagon and regional headquarters but also for operational military units—an essential component of the Revolution in Military Affairs.

Even with the GCCS in place and expanding, and given the implications of converting the U.S. military into a digitized force, it is probable that the U.S. military in the new century will increasingly rely on civilian networks to handle much of its operational communications needs.

Another factor that will govern the transformation of military communications is the strategic assumption (which is unlikely to change) that the U.S. will continue to operate in close concert with foreign allies and coalitions in responding to various emergencies and conflicts.

Our future military communications systems must also be able to pass on and receive information from non-U.S. forces and other sources. This is particularly true in engagements involving peacekeeping, peacemaking, and other contingencies in which we operate as part of a coalition of nations. This means we have to maintain the capability to communicate with less-advanced nations as we continue with the Revolution in Military Affairs. The Defense Department will need to undertake a concerted but careful program to encourage the sharing of advanced technological applications—particularly with communications and command-and-control gear—not only with our traditional allies, such as NATO, but also with ad hoc partners who emerge as part of emergency coalitions to deal with future crises.

The system we are building is complicated and continually changing. It includes wireless (satellite and radio) communications that provide a "backbone network" and a "deployable theater extension network" to serve dispersed users. Some of the satellite resources are reserved for exclusive military use, such as the Defense Satellite Communications System, which carries high-data-rate circuits for high-volume transmissions; the Fleet Satellite Communications System; and the Milstar communications satellite network. The military also uses a variety of terrestrial radio systems, including combat net radios, high-frequency radio networks, and other systems that do not depend on satellites. And we also lease parts of commercial communications systems.

But the promise of a civil-military communications partnership is still a number of years off. What the Pentagon uses today is a collection of very heterogeneous military communications systems with some important limitations. Many of the systems rely on various stages of technology—everything from old-fashioned cathode-ray-tube radios, to early transistor models, to what now seem to be antique PCs running on DOS, to the latest ultrathin laptops—and because we have invested billions of dollars in these systems and spent countless hours training our forces to use them, it seems likely that they will be kept in use for years. Unfortunately, this practice will continue to make communications across military services harder than it should be. Even today, a decade after Operation Desert Storm revealed a number of serious "interoperability" flaws that hampered communications among the four combat services, problems still remain. The "legacy" communications systems maintained by the military services are probably the most important technical barrier to better joint operations and the leading factor delaying the appearance of the digital communications "system of systems." I can't say strongly enough how much the Revolution in Military Affairs depends on our revamping our communications systems.

The problems the military experienced with communications during Operation Desert Storm illustrate the limitations of the current system. Anticipating a surge of messages when the fighting started, the Pentagon moved an additional Defense Satellite Communications System satellite and over a hundred additional ground satellite terminals to the Arabian Peninsula. It was not enough, so the Pentagon supplemented the operation with commercial terminals. But our people were simply sending too many messages; the demand for bandwidth still far surpassed the available circuits.

There has been some progress. The new GCCS network represents a leap ahead and the start of a transition to a different U.S. military communications system.

We sometimes use the metaphor of an information "superhighway" to portray what is emerging in terms of information technology. But that term fails to capture the global "*info*structure" the Pentagon is creating. (Highways have limited capacity and permit entry or exit only at specified points. What we are building has virtually unlimited capacity and will allow access nearly everywhere for all U.S. military forces.) We are leaving a communications system of specialized networks and classical circuit switching for a global communications grid. The frequency spectrum, the growing space-based communications infrastructure, the broadcast media—all of this is better understood as a continuum that provides paths for the transfer of digital packets between U.S. military users anywhere on earth. The new communications system will enjoy virtually unlimited interconnectivity, decreasing the likelihood that local or regional disturbances, jamming, or physical attack could prevent communication between points in the grid.

The new communications "system of systems" will be digital and will support all communications media. In addition to data transmission, it will provide voice and video communications throughout the operational chain of command, from the Pentagon to small combat units in the field—a degree of connectivity that we have not yet seen even in operations as recent as that in Kosovo. At the risk of repetition, it has been our inability to conduct instantaneous and secure communication that has traditionally thwarted the U.S. military commanders' ability to succeed in their combat duties. Because the system will be digital, we will be able to break any information into packets, send the packets to any point in the grid, and reassemble the information there. In effect, the system will be able to pass any information to any user and any user will be able to draw out the precise information that is needed. In effect, this new mode of communication will transform all elements of the U.S. military—from major units to individual soldiers—into sensors. What they see, sense, and know will immediately be added to the senior U.S. military commander's understanding of the battlespace.

We are only beginning to understand what this new communications system means and to recognize the profound significance it may have with regard to traditional concepts of military command, control, hierarchy, and organizational structure. Historically, militaries have apportioned knowledge by rank: The higher the rank, the broader the perspective, the greater the knowledge. When one of Napoleon's soldiers went to the front at Waterloo, for example, all he saw was the back of the soldier in front of him; and Napoleon himself was forced to communicate with his subordinate generals by writing letters of instruction that aides then had to carry on horseback, often taking hours (if not days) to deliver. Throughout the twentieth century the speed and surety of communications throughout the military chain of command has steadily—sometimes fitfully—improved, but the final hurdle remaining to be overcome has been the combination of universal access to communications and instantaneous transmission over digital networks.

The final breakthrough promised by the current Revolution in Military Affairs is to provide the military commander the ability to outperform any enemy by being faster—faster to learn the position of his foe, faster to transmit the correct orders to attack, and faster to assess whether the attack has succeeded or whether subsequent strikes are required. The communications infrastructure of the Revolution in Military Affairs that we can develop within the next decade or two will challenge the historic pyramidal architecture of military command because it will allow and prompt a much broader diffusion of knowledge that is relevant to the instantaneous demands of combat throughout all levels of the armed forces. Indeed, many futurists are studying the radical concept of "netwar" in which smaller, highly mobile but lethal strike forces—interconnected by digital communications—will replace today's inefficient, bulky, and immobile Army, having the capacity to swarm across the future battlespace to strike and destroy the enemy. That is the promise of the Revolution in Military Affairs.

Although there is some evidence that we are moving in the right direction, the struggle to transform military communications indicates that the Pentagon has lost its way. Having continued to closely and carefully watch the military procurement and operating budget since my retirement in 1996, I am convinced we are not moving fast enough. The Pentagon's central nervous system—the myriad layers from satellite terminals to desktop telephones—remains a truly Byzantine structure that can baffle even the most seasoned expert. The plethora of agencies, offices, groups, institutions, and coalitions that shape the U.S. military's ability to communicate often defy comprehension. But the implications of the revolution in communications that is emerging—all too slowly—are profound. We cannot afford to drag our feet. For unless our ability to communicate dominant battlespace knowledge keeps up with our capacity to generate it, the American Revolution in Military Affairs will not succeed.

Part 3: Precision Force and the Commander's Intent

If the U.S. military commander can see and understand a battlespace, and communicate that understanding throughout his combat and support units, he will be well positioned to dominate that battlespace. Militaries dominate by virtue of force and violence. The manner in which they dominate and the risks they incur in doing so are functions of how they can employ violence. This is one more of the factors that is increasingly distinguishing the U.S. military from that of other nations, for we are increasingly able to bring force to bear faster, over greater distances, with precision and accuracy, and—as we have seen in the Gulf War and the war in Kosovo—with a far smaller risk to our fighting men and women. At the most general level, the United States is entering the era of precision violence long before other nations.

Precision guidance refers to a class of munitions distinguished by their accuracy and the ways in which it is obtained. Today there are three general categories of guidance—all of which have generally become known to the public as a result of the Pentagon's release of video images of their use in conflicts ranging from Operation Desert Storm to Kosovo.

The first category still requires human movement—directly or indirectly—to guide the weapon once it has been launched, fired, or dropped. Laser-guided bombs, or anti-armor missiles like the TOW missile (for Tube-launched, Optically-aimed, Wire-guided), require that a human either keep his eye on the target or train a steering beam, often a laser, while the weapon approaches the point of impact. This kind of precision-guided munition has the advantages and disadvantages you'd expect. We have yet to devise an electronic sensor with the sophistication of the human brain, but with this guidance system, the weapon will lose accuracy and miss its target if the shooter becomes distracted while directing the missile or bomb to its target.

The second category of precision-guided munitions consists of more autonomous weapons. These weapons carry their "smarts" with them in the form of an electronic sensor that searches for the target and, upon locating it, guides the munition to impact. The sensors can utilize infrared radiation (the Sidewinder and Stinger antiaircraft missiles are good examples), acoustic energy (modern antiship torpedoes), or radar (most long-range antiaircraft missiles and almost all antiship missiles). Some weapons use a combination of different sensors such as the new Brilliant Antitank (BAT) munition, which when fired from an artillery tube initially searches the target area by listening for certain acoustic frequencies that alert it to the location of the vehicles it hunts—and switches over to an onboard infrared sensor to guide it to impact. Early versions of the Tomahawk cruise missile carried preprogrammed maps showing the route to the target, which the missile would "check" by comparing signals from terrain-mapping radar in its nose; then when close to the actual target the cruise missile would compare an optical image of the passing landscape with a recorded picture of the target.

The third category of precision-guided munitions consists of relatively autonomous "fire and forget" weapons that rely on externally generated information to guide them to their target. More recent models of the Tomahawk cruise missile and the air-dropped Joint Direct Attack Munition (a guidance kit that converts a 2,000-pound gravity bomb into a "smart" weapon) contain an onboard computer that literally flies the weapon to a specific point on the earth's surface by comparing onboard data against the data provided by the Global Positioning System (GPS) navigational satellites. The heavy reliance on these latter two weapons during the seventy-eight-day air campaign in Kosovo in 1999 demonstrates the Pentagon's awareness that the other types of precision weapons would be insufficiently accurate to avoid accidental civilian casualties. Currently these point-guided munitions are targeted against fixed targets, but soon they will be able to receive real-time data updates from battlefield sensors so the munitions will be able to hit moving targets.

Where should precision-guided munitions have their brains? Munitions that carry their own seekers are relatively complex and expensive, and the more they cost, the more difficult it is to buy them and keep them up-to-date. One reason for the high cost of the Kosovo air campaign stemmed from the high price of the advanced precision-guided weapons that were used. During the eleven weeks of air strikes, Navy ships fired about 450 Tomahawk cruise missiles, at a price of about $1 million a missile. Meanwhile, Air Force B-52 bombers fired 90 air-launched cruise missiles, which cost about $2 million apiece. And the B-2A stealth bombers dropped hundreds of Joint Direct Attack Munitions smart bombs, forcing the Pentagon to reopen the production line as stockpiles plummeted.[29]

These new munitions are more vulnerable to disruptions in GPS signals and the off-board targeting communications they depend on for their accuracy; yet because they are drawing on the information provided by dozens of powerful external sensors rather than just a few onboard ones, they can lock onto even a vaguely defined target—say, a low-flying helicopter or ground vehicle—and follow suggested tracks rather than having to go to where the target last revealed itself. Since they don't have to carry their own sensors, point-guided munitions are cheaper—which makes targeters prefer using them against less valuable targets or in large numbers to get through by saturating defenses.[30] To the extent that we achieve the promise of the ISR and C^4I technologies described earlier in this chapter, point-guided munitions will become increasingly appealing to U.S. military planners.

The technological keys to these weapons systems lie in the NAVSTAR Global Positioning System, the precise geo-coordinate satellite constellation, which allows us to accurately determine the position of objects in terms of latitude, longitude, and elevation. The NAVSTAR system is central to the "smart" war of the present and is likely to be even more important in the future, so it's worth explaining it in a bit more detail.

The GPS network uses a constellation of twenty-four satellites in twelve-hour, 12,500-mile-high orbits. The orbits are precisely defined, and at any instant each satellite's actual position can be defined to within 3 meters. In effect, the position of each satellite serves as a fixed base from which to locate any other object. By measuring the distances to two or more of the satellites, it is possible to accurately calculate the precise location of any point on or over the earth in terms of its latitude, longitude, and elevation. The measurements are made based on radio signals transmitted from the satellites; each satellite transmits low power signals. Specially designed equipment can receive these signals, lock up to them, and track the code and carrier phase of the signal.[31]

Since in a future conflict it is likely that an opponent with advanced military capabilities would attempt to jam the GPS signals, the U.S. military has designed ways to get around this kind of interference. For example, it is possible for the Pentagon to supplement GPS-based navigation with inertial guidance devices that "fill in" for gaps in received GPS signals. Or the GPS system itself can use its capabilities to determine the location of the device that is doing the jamming, providing U.S. forces with the information they can use to destroy it in retaliation. (Only the United States and the former Soviet Union ever actively researched the possibility of fielding antisatellite weapons that could directly threaten the satellites themselves.[32])

It will also be possible to locate any object on or above the surface of the earth with great accuracy. A combination of electro-optical imaging satellites and advanced synthethic aperture radar have provided us precise maps of the surface of the earth. With this knowledge (supplemented by differential GPS and other techniques), we can locate any object to within about 1 meter of its actual location. In other words, we will know precisely where we are and where any identified target is located. This means we will be able to strike anything we see and can reach with a smart weapon. I believe that within the very near future the U.S. military arsenal will field point-guided weapons that will use real-time updates to home in on moving targets.

These innovations will mark the crossing of an important threshold on the path toward applying force with near-perfect accuracy and precision. Beginning shortly after World War II, electronic advances enabled sensor-guided weapons (such as television-guided bombs) to achieve high accuracy—within about 10 meters. That was more than an order of magnitude better than the average accuracy of the bombs delivered in the war. But seeker-guided weapons have some important limitations. The target must have particular characteristics—the right "signature"—for the seeker to home in on it, and the weather conditions and attack attitude must allow the sensor in the weapon to acquire the target at sufficient range to permit guidance. Other kinds of relatively longer range precision-guided weapons—such as laser-guided bombs or rockets—require the continued involvement of the launching platform or other platform, usually carrying a human, until the weapon reaches impact. This arrangement increases the vulnerability of the platform carrying the designator. And sensor-guided weapons are usually expensive. Point-guided weapons capable of navigation accuracies of 1 meter or better will be low-cost,[33] all-weather, launch-and-leave weapons that can be used at great standoff ranges.

That gain in accuracy alone will significantly change the way we think about combat and how we wage war. High accuracy will enable much smaller warheads to get the desired level of destruction. The combination of a small radius of lethal effects (from the smaller warhead) and the low (or zero) miss rate means less collateral damage. If the weapons are smaller, the platform that carries them can carry more, and the more it can carry, the more targets it can destroy before reloading. Alternatively, smaller platforms armed with smaller weapons can destroy relatively more targets than they could were the platforms armed with less precise, larger weapons. Increased effectiveness at greater ranges will reduce the vulnerability of the platform that launches the weapons (and the men and women who operate it). The increased combat effectiveness of the weapons will increase the operational tempo of combat and reduce the size of the logistics support pyramid on which operations are based.

One aspect of the Revolution in Military Affairs that has been encouraging to me is the Pentagon's steady progress in fielding new technological applications that reinforce the core military imperatives discussed in this chapter—the commander's ability to "see," to "tell," and to "act."

The most persistent disappointment I've experienced is to discover over and over how little anyone in senior military command in the United States has grasped the whole potential of the Revolution in Military Affairs. Consider the list of acronyms in Table 3-1 (see p. 148) that are components of the "system of systems."

As vice chairman of the Joint Chiefs of Staff, I began showing this list to various military audiences consisting of senior commanders and experts on military technology. I could always find someone who knew more than I thought was humanly possible about one or two of the systems listed. But never—and I emphasize the point, *never*—could I find a general, an admiral, or a senior civil service employee (including myself) who understood in detail and in depth what they *all* were, or, more important, what *all* of the systems meant *together*. The most

Table 3-1

U.S. Military Systems That See, Tell, and Act

ISR*	C^{4}I†	Precision Force
AWACS	GCCS	SFW
Rivet Joint	MILSTAR	JDAM
EP-3A	JSIPS	JSOW
JSTARS	DISN	TLAM
HASA	SABER	ATACMS/BAT
SBIRS	C^{4}IFW	CALCM
Tier 2 (+)	TADIL J	HARM
Tier 3 (–)	TACSAT	THAAD
TARPS	TRAP	Hellfire II
MTI	MIDS	Javelin
UGS	SONET	LONGBOW
ISAR	etc.	LOSAT
Magic Lantern		HAVE NAP
Predator		AGM-130
NVG		etc.
Cobra Ball		
etc.		

Note: A full listing identifying these elements and their functions can be found in the Appendix on page 239.

*Intelligence collection, Surveillance, and Reconnaissance.

†Command, Control, Communications, Computers, and Intelligence.

important word in each column is "etc." because there are actually *hundreds* more elements in each category. Since they cost our nation hundreds of billions of dollars, our leaders should know more about them!

What is possible today is a vast new "system of systems" that is based on the integration of many components. It is the combination of those elements that will give the "system of systems" its full strength and make possible the full success of the current Revolution in Military Affairs.

We have begun to recognize that the "system of systems" that promises so much is not simply composed of new tanks, aircraft, or ships—the sort of tangible assets that are normally addressed in balance assessments of contending militaries. We perceive—still too dimly and too slowly—that the secret strength of the components of the "system of systems" has more to do with what these assets carry and the technologies they employ that can help weld tanks, aircraft, ships, and other major components of military strength into a more effective, efficient, deadly force.

CHAPTER 4

LAUNCHING THE REVOLUTION

I can pinpoint the moment when I realized how dysfunctional and inefficient the U.S. military had become, and how sweeping was the need for a thorough reorganization of the armed forces to meet the challenges of the new century. It came midway through my tenure as commander of the 6th Fleet in the Mediterranean, when, several months after the end of the Gulf War in 1991, I decided to try an experiment in emergency communications. For this exercise I told each of my naval fleet units that I wanted it to demonstrate that it could communicate directly with a U.S. Army ground combat unit—any combat unit—somewhere in Europe. It is, of course, possible for U.S. ships to communicate with U.S. ground forces, and we do so all the time. But the means of communicating are far from direct. Complex, sometimes convoluted routings and different communications channels, switchboards, and operators are involved. For this experiment I wanted the communication to be direct from a 6th Fleet naval unit at sea to the Army unit ashore in Europe—not routed through a switch at some communications center or higher headquarters.

We tried to do it. We tried—*for six months*—to make the connection. And we never could link up with the Army. There was always something wrong. The Army and Navy units used different frequencies. The communications protocols employed by the Army and Navy were different.

The communications personnel had never been trained to do this. And so on.

In mid-1992 I was posted back to Washington, D.C., to serve on the newly reorganized Navy staff under Chief of Naval Operations Admiral Frank Kelso. I could not get out of my mind the frustration I had felt following the fruitless ship-to-shore communications exercise. What was it about our military services—and the Pentagon's leadership—that would allow such a serious problem to go on indefinitely?

I was beginning to believe that military parochialism—defined at the individual level as a service member's "traditional" loyalty to service or military specialty over the armed forces as a whole, whatever his or her rank or position—would probably turn out to be the most serious obstacle preventing meaningful reform of the Navy and the other services so that our combined services can adapt to the new world around us.

The peacetime exercise was frustrating and maddening to all hands involved. But the consequences of such a mismatch in war are much more serious. It can be deadly and devastating if a combat emergency occurs requiring such a communications linkup and that connection cannot be made.

THE EFFECTS OF MILITARY SERVICE UNILATERALISM

The inability of two different military services to communicate directly with one another is characteristic of the wasteful, inefficient, and potentially dangerous rivalry that continues to undermine the military today. Worse, it is a problem that has haunted the armed services for most of the last century. From Pearl Harbor in 1941 through Kosovo in 1999, the United States has struggled with the consequences of this problem: botched military operations, "friendly fire" incidents in which we accidentally shoot at our own troops, and "collateral damage" where our military violence accidentally kills innocent civilians.

Poor training, inexperience, and the inevitable stresses of combat are common grounds for such mishaps, and a flawed policy or decision by the National Command Authority—the President, the secretary of defense, and the chairman of the Joint Chiefs—can lead to military disaster before the first shots are fired. But as I see it, the root cause of most of our recent friendly-fire incidents is the deep-seated isolation of the individual U.S. military services from one another. Today, when "synergy" and "strategic alliance" are common parlance for every middle manager in the U.S. business community, our Army, Navy, Air Force, and Marine Corps essentially refuse to communicate with each other. In fact, they compete more than they cooperate: and what's worse, they defend this approach as a noble tradition that is crucial to their history and their esprit de corps. The Pentagon's adoption of recent policies encouraging "joint" doctrine and interservice cooperation has had scant effects.[1] The massive military structure we retain today is itself the product of what I call *military service unilateralism.* Despite all the progress reports on multiservice cooperation and "jointness," the four combat services still operate within an organizational structure reflecting decades of bureaucratic rivalry. A list of the worst mishaps stemming from this wrongful competition strongly resembles a thumbnail chronology of U.S. military history:

- *Pearl Harbor, December 7, 1941.* Following the surprise Japanese attack on the Pacific Fleet that pitched the United States headlong into World War II, nearly half a dozen major investigations into the disaster found a central theme of Army-Navy mistrust and lack of communication between the two services on Hawaii as a major contributing cause for the failure to conduct aerial patrols and place combat units on alert despite clear warnings from Washington that the Japanese could attack imminently anywhere in the Pacific. As a result, 2,403 U.S. servicemen were killed, 18 warships sunk or heavily damaged, and three-fourths of the Army air corps aircraft on Oahu destroyed.[2]
- *Operation Cobra, France, July 1944.* Six weeks after the successful landings of Allied troops at Normandy, Allied commanders decided to organize a massive bomber raid on German defensive lines to blast an opening into the French interior for the ground forces. Planners immediately disagreed over how close to the U.S. front lines the bombers could safely drop their bombs. The Army insisted that the strikes should occur no more than 800 yards away, but the

air commanders said the minimum safe distance was 3,000 yards. The ground force commanders also insisted that to ensure the safety of ground troops the bombers fly from north to south so that their bombs would fall parallel to the American positions, but the air commanders argued that the aircraft should fly from west to east to minimize the planes' exposure to German antiaircraft fire. The debate foundered on mutual misunderstandings on each side about the other group's combat doctrine: the ground commanders wanted the minimum safe distance of 800 yards because they feared the Germans would retake the target area before our ground troops could attack; the air commanders feared (correctly) that they could not rule out bombs falling astray. On two successive days a force totaling 1,800 bombers plastered the front lines, killing 111 Americans (including General Leslie McNair, at that point the only four-star general killed in World War II) and injuring another 538. Poor communications between the ground force commanders in France and the air force leaders back in England had led to catastrophe.[3]

- *The USS* Pueblo, *February 23, 1968.* The North Korean seizure of the electronic surveillance ship USS *Pueblo* in the Sea of Japan revealed a colossal breakdown in communication between the U.S. intelligence community (particularly the National Security Agency, which directed the ship's surveillance operations) and the U.S. Navy, which was responsible for administrative management of the *Pueblo* and its crew. As a result, U.S. military commands in Japan and South Korea were unaware of the ship's mission and no forces were on alert and close enough to come to the defense of the *Pueblo* once it came under attack.
- *Desert One Site, Iran, April 25, 1980.* The U.S. commando operation to rescue 52 American embassy hostages in Iran was organized under extreme secrecy in order to avoid giving advance warning to the Tehran revolutionaries. But following the mission's abysmal failure at the initial landing site inside Iran, where a helicopter collided with a C-130, killing 8 U.S. servicemen, an official probe into the operation found serious, systemic flaws in the planning and organization of the raid. Because of secrecy most of the subordinate commanders were unaware of the plan as a whole, and prior to the mission there was no review of the plan, and thus no one

detected the serious flaws in the operation's design. Critically important, the probe found that the key team of Marine Corps pilots responsible for flying seven RH-53 helicopters from the USS *Nimitz* to the Desert One Site were never properly trained and could not even communicate with each other, with the ground forces, or with the Air Force aircraft (the Marines were flying Navy helicopters with radio sets that were different from what they normally used). As a result, several helicopters failed to reach the site, forcing the commander to abort the raid. And miscommunication between the pilots and the airstrip was a major reason for the collision that followed.[4]

- *Grenada, October 27, 1983.* Obscured by the blackout of press coverage that hindered in-depth reporting on the invasion for months, Operation Urgent Fury is seen by experts today more as an indictment of poor interservice communication than as an inevitable military victory. Despite the engagement being designed as a multiservice operation, Army ground forces could not communicate with their Marine Corps counterparts even though they were supposed to be conducting closely coordinated missions on the island. Nor could Army soldiers on the ground talk directly with Marine Corps or Navy aircraft. At one point, according to press accounts, an Army officer in urgent need of air strikes from Navy aircraft overhead could see the carrier USS *Independence* offshore but could not reach the ship with his Army radio. In desperation, he used his personal telephone credit card at a pay telephone to call his division operations center at Fort Bragg, North Carolina, which then relayed the message to the Navy, which routed the message to the carrier USS *Independence*, which then contacted its combat aircraft over Grenada.[5]
- *Northern Iraq, April 14, 1994.* After being vectored and cleared by a U.S. AWACS E-3B aircraft, two U.S. Air Force F-15 fighters patrolling the no-fly zone over northern Iraq fired on two unidentified helicopters flying on a straight-line course at relatively low altitude. It was a textbook engagement. Both helicopters were destroyed on the first F-15 pass. Tragically, both helicopters were U.S. Army Blackhawks carrying 26 military and civilian officials involved in the ongoing Operation Provide Comfort to aid dis-

placed Iraqi Kurds in the region. The Pentagon investigation concluded that the incident was an "avoidable tragedy" that had occurred as a result of crew error and a possible flaw in the "Identification, Friend or Foe" transponder transmitters aboard the helicopters, which could have prevented the helicopters' systems from sending the correct signal to the F-15s identifying the helicopters as belonging to the U.S. Army. But once more, probers found poor coordination between Army and Air Force flying units; it was common for Air Force planes to have no knowledge of Army helicopters' whereabouts.[6]

- Friendly-fire casualties also marked one of the most serious failures of the U.S. military during Operation Desert Storm, during which 24 percent of the American fatalities—35 service personnel—were killed by friendly fire. There has been little in-depth discussion of this issue in the decade following the Persian Gulf War, but the facts still command our attention. As a congressional report in June 1993 found that "there were 615 American battle casualties—personnel killed or wounded—in Operation Desert Storm, 148 of which were fatal. Of the 148 fatalities, 35—or 24 percent—were caused by friendly fire. Of the 467 nonfatal battle casualties, 72—or 15 percent—were caused by friendly fire. These percentages seemed high at the time when compared to those assumed from wars past but the review of past rates of fratricide suggest that there has been a substantial underappreciation of the rate of fratricide in past wars."[7]

What remains surprising to me is not the "discovery" by experts that the overall rates of fratricidal killing in previous wars have been relatively constant, but that friendly-fire casualties have been so broadly accepted as an inevitable occurrence stemming from "the fog of war." Although there is no question that fratricide has been a tragic part of warfighting throughout the history of man, even a decade ago technology existed to give us the ability to do something about it. The fact that we have failed to take advantage of this know-how raises a question about whether the "management" of the Pentagon—and I include myself—may deserve significant responsibility for this persistent, unnecessary loss of life.

Many friendly-fire deaths occurred when forces of one U.S. military service attacked elements of a different service.[8] This was the case with a majority of the 9 fratricidal incidents in which U.S. aircraft mistakenly struck American ground forces operating in the same sector of the Desert Storm theater of operations. Having reviewed the reports on fratricide, I have concluded that the connecting link in the vast majority of the incidents involved misperceptions and human error caused by the lack of proper data links, the lack of common operating procedures among the different services, and—overall—the residual impact of individual service cultures that precluded a truly "joint" situational awareness of the battlefield.

Friendly-fire incidents are not a theoretical issue. They strike to the heart. One night in 1995, while I was serving as vice chairman of the Joint Chiefs of Staff, I had dinner with twenty sets of parents of Americans who *we* had killed in Operation Desert Storm. It was one of the most difficult days of my entire life. During that dinner I found there simply were no answers to the many questions these parents asked me. Most American parents have some understanding that their children could choose to serve in the military. It is always understood as a possibility, however remote, that their child might even lose his or her life on behalf of the country. But it is something else to confront parents with the fact that their children were lost because fellow U.S. troops unintentionally and wrongly fired on them. The parents' questions to me that night aimed at a central query: Why couldn't we have prevented this? Doesn't American technology provide a solution that could overcome the chaos and fog of war to safeguard them? I had no ready answer on that bleak night. One purpose of this book—perhaps the most important—is to underscore the urgent need to improve battlefield awareness so that such tragedies will not occur in the future.

THE PLAGUE OF PAROCHIALISM

To most civilians, the military is a strange, homogeneous subculture in American life whose inhabitants speak in tongues, wear odd clothing with ribbons and badges of unknown significance, and live according to obscure traditions and laws. This is an illusion; but what most civilians do not know is that the appearance of partnership and cooperation among the four U.S. armed services is the biggest illusion of all.[9]

As I noted in Chapter 1, it's well known that each service has its own organization architecture built on historical experience, traditions, myths, and icons. Most of these distinctive features are quite benign and, in fact, contribute to the esprit de corps and sense of community that is essential to foster service morale and unit cohesion.

And there are sometimes valid reasons for the forces to do things differently. Navy jet fighters must be designed and built differently than their Air Force counterparts, because the physical realities of catapult takeoffs and arresting-gear landings on an aircraft carrier flight deck require a much stronger fuselage and landing gear, as well as stronger protection from salt and corrosion. Helicopters used by the U.S. Special Operations Command must have a different suite of navigational equipment than those used by regular Air Force or Army units because the operational environment in which they fly demands the ability to fly longer, at night, and at low altitudes. The Air Force and the Navy stock gravity bombs that seem to be essentially the same other than the paint. But on the Navy bombs the differences in the paint are necessitated by different safety and storage considerations, which in turn are rooted in the fact that the Navy ordnance is stowed aboard ships in relatively cramped storage magazines, very close to other ship components. Because of safety considerations the Navy has to be particularly sensitive to the flammability of the paint on its bomb inventory, while the Air Force, accustomed to more dispersed storage for its bombs, does not.

Some defenders of the status quo will point out that there is actually very little pure duplication or redundancy across the military service boundary lines. What looks like redundancy is actually useful and necessary specialization.

But in many cases the differences are too subtle and too contrived to be justified, or they stem from each service branch's desire to avoid depending on the others. This separation of forces was not a great problem during the era of industrial-age warfare. But in the information age, when technological advances have blurred the traditional boundaries in space and time that long physically separated the Air Force from the Navy, and the Army from the Marine Corps, the military services' inability to communicate with one another and their indifference to doing so have set the stage for crisis and disaster.

My own service, the Navy, is as guilty of this as any other. Throughout the Cold War, the Navy designed, sized, and shaped its combat units to fight the Soviet Navy in the deep ocean. While naval units operated throughout the world in support of U.S. national interests in engagements as diverse as war in Vietnam and peacekeeping in Beirut, the prospect of a confrontation with the Soviet Red Banner Fleet dominated the sea service's doctrine, planning, force procurement, and day-to-day training missions. By the mid-1980s, during the Reagan Administration's program to modernize and expand the Navy to a 600-ship force, this approach had crystallized in "The Maritime Strategy," unveiled by then Secretary of the Navy John F. Lehman Jr. and Chief of Naval Operations Admiral James D. Watkins in 1986.

The strategy called for large-scale naval operations centered on aircraft carrier battle groups consisting of a nuclear-powered *Nimitz*-class carrier, two or more guided-missile cruisers, a dozen smaller destroyers and frigates, several combat support ships, and one or two nuclear attack submarines. U.S. Navy operations along the periphery of the Soviet Union from the Barents Sea to the northeast Pacific would pose offensive threats to the Soviet Union—by both manned aircraft and the Tomahawk land-attack cruise missile. They would also threaten Soviet strategic missile submarines operating in heavily protected "bastions" near the Soviet coast. Therefore, the Soviet Union would be forced to contest the U.S. Navy for control of these flank areas and thus divert considerable military assets away from the Western European front. Thus the Maritime Strategy sought to contribute to the NATO alliance defense in Western Europe, even though the U.S. Navy would be operating thousands of miles away.

An assumption buried in the logic of that strategy was that the Navy would operate in isolation from the other services. If the Navy was going to operate hundreds or even thousands of miles away from where the Army and Air Force were going to be waging war, the Navy had to depend on its own weapons, equipment, strategies, and personnel.

So the Navy built and deployed air-defense systems specifically designed to defend the fleet in the open ocean in situations where the only aircraft in the sky were likely to be those launched by U.S. aircraft carriers or the waves of Soviet land-based Backfire bombers. The Navy developed and deployed the most sophisticated antisubmarine warfare systems in the world and trained U.S. submarine crews in tactics and strategy to defeat the larger Soviet submarine force. The Navy optimized its radar, communications, and attack systems for a conflict environment in which there would be only two contenders, both very strong, very deadly, and fully committed to killing each other.

So it was that the Navy's objective came to be having the ability to fight and win on its own—rather than to be geared to contributing to an overall U.S. military strategy or a joint battle plan. Meanwhile, the other services were pursuing similar strategies in isolation from the others. The Air Force prepared for strategic bombardment and missile attacks against Soviet targets. The Army concentrated on its ground defense against a Soviet invasion of Western Europe and threats by other hostile powers such as North Korea. And the Marine Corps—a smaller service much of whose history has consisted of defending its very existence against the threat from the rest of the U.S. military—consistently battled not only to preserve its amphibious warfare mission but to retain and protect its own organic naval (transport) and air arm.

Rooted in the limits of military technology in the pre-computer era, and nourished by bureaucratic rivalry permitted by a command structure that gave no one military leader the final say in any major decision, *military service unilateralism* ruled the U.S. armed forces throughout the Cold War. Duplication and redundancy among the different services became the norm, and once in place, redundant programs became heavily vested. Each service worked hard to ensure that it received the funding and operational priority, while working hard to defeat the other services if they developed a rival capability.[10]

Historically, the military services' strongest argument for the duplication of effort and widely overlapping capabilities has been that the friction and fog of war require them to compensate for the unexpected and surprising events that crop up in combat.

Yes and no. We have enjoyed military victories but the costs of this unilateralist approach—in terms of blood and money—have been much higher than they ever had to be. And we have suffered embarrassing failures, as noted earlier. But today the technology exists to complete the Revolution in Military Affairs and to transform the way the United States organizes its armed forces to conduct warfare. Doing so will challenge us to rethink the military traditions and cultural roots that prevent us from taking full advantage of that technology in a fully integrated military force.

A HALF CENTURY OF MILITARY GRIDLOCK

To begin attempting a solution, it's important to recount the recent history of how our military developed its current shape and structure. This history began after World War II, the war that defined our modern military, particularly the deeply held belief in the autonomy of each individual service to decide how it will fight, and how it will organize, train, equip, and maintain its combat forces.

World War II is sometimes exalted as an event that drove American military services toward joint operations. This interpretation is only partly accurate. It was during World War II that military technology matured to a point where the service branches began to overlap, often grudgingly and in an ad hoc way after a disaster had occurred. The Battle of the Atlantic in early 1942 is a prime example. When the United States declared war on Germany, the German U-boat force immediately began ravaging commercial shipping along the East Coast of the United States, and our Navy found itself with insufficient numbers of escort warships and patrol aircraft to counterattack the enemy submarines. But for two months, then Chief of Naval Operations Admiral Ernest J. King resisted all efforts to implement shipping convoys and to use either civilian small boats or combat-effective Army patrol planes to fill in the gap in the coastal antisubmarine defenses, a decision that one pair of military historians termed King's "greatest misjudgment" in the entire war.[11]

As the war continued, operational necessity usually overcame service parochialism, and King accepted the use of ground-based Army patrol planes. But this particular "friction" between the sea and ground services was neither inevitable nor warranted under any philosophy, especially in a worldwide conflict lasting forty-four months and involving nearly 9 million servicemen and women whose service roles and missions would inevitably overlap. The Army built a ship inventory so large that it rivaled the U.S. Navy, while the Navy Department, through the U.S. Marine Corps, built a land force of nearly 500,000 that rivaled the size of the U.S. Army ground force at the invasion of Normandy. The U.S. Army Air Corps, fighting for a separate postwar identity, looked nervously over its shoulder throughout the conflict as the U.S. Navy boosted its naval aviation arm to a total of 40,000 carrier- and land-based aircraft.

The rivalry in the Pacific betwe en Army forces led by General Douglas MacArthur and Navy forces led by Admiral Chester W. Nimitz is a prime example of how the U.S. political and military leadership allowed the grand strategy of the war to hinge on military service prerogatives and not on the logic of a grand strategy to defeat the Japanese empire.[12] Rather than make the hard decision to select one overall commander, the Pentagon allowed the two separate U.S. forces to operate separately. The direct descendants of these forces today are the Pentagon's nine regional military commands and the four-star Army, Navy, Air Force, and Marine commanders who serve under them directing multiservice operations worldwide today.

When the war ended, President Harry S Truman sought to rein in the individual services, issuing a plan to create a single Department of National Defense, but his effort foundered largely because of strong opposition by the Navy.

In its place, Congress passed the National Security Act of 1947 and, by 1949, had created the Department of Defense and the Joint Chiefs of Staff with a chairman as titular military chief. Even so, the individual services retained autonomy because the chairman of the Joint Chiefs did not—and could not—formally command the service chiefs or their separate fiefdoms. The artificial division of combat missions and bureaucratic turf was formally decided at a conference in Key West, Florida, where then Secretary of Defense James Forrestal and the Joint Chiefs specified the primary and collateral functions of each service.

Not surprisingly, the Army would be responsible for ground operations; the Navy for sea operations; the Air Force for air operations; and the Marine Corps for amphibious operations. Although the services were supposed to coordinate with one another on "all matters of joint concern," the independence of each service was clearly retained. And most important, the services *controlled their own individual budgets*!

This organizational scheme formally established the precedent for decades of bureaucratic rivalry among the four combat services over roles and missions and their share of overall defense spending.[13]

When Robert McNamara was appointed Secretary of Defense by President Kennedy in 1961, he became the first civilian leader at the Pentagon to attempt to establish civilian control over the military through its planning and budget process. As president of the Ford Motor Company, McNamara was credited with saving the company from economic ruin through a rigorous program of systems analysis. In his new role he sought to seize control of the Pentagon budget process in a similar way.

The McNamara plan rested on two important assumptions. The first was that there was a general consensus across all the military services as to the primary national security threat to the United States: the growing nuclear and conventional military power of the Soviet Union. Each military service, in McNamara's view, might see the specifics of the threat through the lens of its own unique perspective, but they all agreed on the central premise. The second assumption was that regardless of the consensus, no military service would sacrifice funding for its core mission to accommodate increased "joint" capabilities. Moreover, McNamara's plan imposed complex new procedures on the individual military services in the form of the planning, programming, and budget system (PPBS), which constituted a detailed programming plan that shifted the initiative on major budget decisions to him and his staff. Although critics charged that the PPBS constituted an additional layer of bureaucratic interference in traditional military prerogatives, the use of "systems analysis" by McNamara's staff (whom career officers mocked as "McNamara's whiz kids") did serve as a tool for reaching rational decisions over controversial and expensive military projects.

But McNamara's initiative never really challenged the military services' role in setting defense requirements and spending. (And to this day, the budgets that are submitted by the military services are changed by a mere percentage point or two by the civilian leadership or their military superiors on the Joint Chiefs of Staff. Budget autonomy remains the key element of service parochialism and independence.) Nor did McNamara ever succeed in preventing the services from appealing directly to their patrons in Congress for funding. And he never made a dent in how the services organized themselves or developed combat doctrine. Indeed, McNamara's challenge to the military was not taken seriously, because, in spite of his military service during World War II, he was seen as a man who had never been initiated into the "brotherhood of arms."

For more than a decade after McNamara's tempestuous reign at the Pentagon, his successors as defense secretary focused on the twin issues of the military's recovery from its disastrous Vietnam involvement and its challenge by a growing Soviet military force. The root problem of interservice rivalry and budget autonomy receded into the woodwork.

The reassertion of various military service prerogatives reached a post–World War II high during the Reagan Administration from 1981 to 1988. Responding to the administration's eagerness to build up U.S. military power and its belief that the military itself understood its own needs best, each of the individual services was able to control—more than ever before—the defense secretary's formal defense guidance document for that service's own military programs, and the civilian bureaucracy in the office of the secretary of defense retreated from that responsibility. By the mid-1980s the military services fully dominated the Pentagon budget process.

But responding to several military mishaps in the early 1980s, the Senate Armed Services Committee in 1985—strongly supported by a growing Military Reform Movement in the defense community—published a lengthy staff report criticizing the military for failing to improve its "joint," (interservice) cooperation in several major operations. The outcome of this movement was a series of congressional hearings that led to the passage of the 1986 Defense Reorganization Act spearheaded by Republican Senator Barry Goldwater and Representative Bob Nichols, both staunch supporters of defense.

The new law centralized military authority in the chairman of the Joint Chiefs of Staff in several key areas. First, it made the chairman the principal military adviser to the President and the secretary of defense. Second, the law also created the position of vice chairman of the Joint Chiefs of Staff to provide the chairman with a senior deputy responsible for assisting in the implementation of joint cooperation and in the management of the armed forces. Third, the legislation transferred the Pentagon Joint Staff from the committee-like entity of the Joint Chiefs themselves to the office of the chairman, providing the chairman with a powerful tool to carry out his leadership functions. Fourth, the law significantly strengthened the power and influence of the regional commanders in chief (CINCs), making them the principal operational commanders of the U.S. military reporting directly to the secretary of defense (with the defense secretary having the prerogative to designate the chairman of the Joint Chiefs as a communications link with the CINCs).

The congressional intent behind the Goldwater-Nichols Act was to make "jointness"—the formal concept of interservice cooperation and planning—the law of the land. Service on a Joint Staff, long regarded (accurately) by officers as a career-ending assignment, now became mandatory for advancement to senior rank.

The 1986 law constituted the most radical shift in Pentagon organization and delegation of power since the establishment of the Department of Defense thirty-nine years earlier. It sparked the first open debate on previously sacrosanct subjects such as overlapping military service missions and the wasteful redundancy of duplicated programs. And it created a new environment within the Pentagon, permitting decision makers to go beyond rhetoric and initiate real reforms for the U.S. armed forces.

As noted later in this chapter, the full powers accorded the chairman of the Joint Chiefs would not be exercised until Army General Colin Powell arrived as chairman in 1989. But as I'll discuss later, Goldwater-Nichols was only a beginning.

TRANSFORMING THE NAVY

The Navy was in a state of siege in the summer of 1992 when I reported to the office of Admiral Frank Kelso, chief of naval operations, to take up my new job as the first deputy chief of naval operations for resources, warfare requirements, and assessments, the senior three-star position on the Navy staff. It seemed hard to grasp that just six years earlier, in 1986, then Secretary of the Navy John F. Lehman Jr. had won congressional funding for two *Nimitz*-class nuclear aircraft carriers in his plan for a 600-ship Navy centered on 15 aircraft carrier battle groups and 100 nuclear attack submarines. The celebrations had not lasted long. In early 1988, just months after both Lehman and Secretary of Defense Caspar Weinberger left office, a new defense secretary and a new Navy secretary were locked in a furious debate over the future size of the service. James Webb, Lehman's successor as secretary, who had distinguished himself in combat in Vietnam as a junior Marine officer, angrily resigned when pressured by Secretary of Defense Frank Carlucci to trim the size of the fleet to conform with planned defense spending reductions. Over the next four years the Navy budget plummeted, and it appeared inevitable that the number of ships in the fleet was going to be less than 450, even though the fleet size held tightly to 600 (that was then; at present we anticipate that the fleet will consist of fewer than 300 ships).

Even with the cuts in warships, the Navy was simultaneously grappling with a personnel shortage that strained our ability to deploy fully manned crews, and a surplus of shore bases that threatened to drain vital budget funds. To make things worse, several major new aircraft programs earmarked for the carrier aviation arm had been summarily canceled because of major cost overruns and mismanagement, particularly the A-12 Avenger stealth bomber and the P-7 land-based patrol aircraft.

And those were merely the *external* forces buffeting the Navy. In the eighteen months after Operation Desert Storm, many senior Navy leaders had come to the conclusion that the service had been ill-prepared for the Persian Gulf conflict. For in Operation Desert Storm the Navy found itself elbow to elbow with the other U.S. combat services and our coalition allies—and unable to work with them very effectively.

Operation Desert Storm also revealed that the Navy's Cold War-era doctrine—which focused the force's tactics, training, weapons procurement, and operational deployments for combat against the Soviet Navy in the deep ocean—was obsolete; naval commanders suddenly had to improvise new tactics and combat procedures to fight the Iraqis without inadvertently endangering either their own aircrews and ship crews, or their comrades in the other services.

The second shock was the impact of the Base Force concept unveiled by General Colin Powell, chairman of the Joint Chiefs of Staff, and Secretary of Defense Dick Cheney in August 1990 in an initial response to the thawing of U.S.–Soviet relations and the ending of the Cold War. Navy leaders had assumed that the Air Force and Army would bear the majority of the force reductions and budget cuts since the reorganization included large-scale Army and Air Force base closures overseas, and substantial unit deactivations, particularly in Europe. In 1990 the senior admirals assumed that because the Navy and Marine Corps would still be required to deploy and serve overseas, their share of the falling Pentagon budget would be relatively higher than that of the other services. And they were dead wrong. In early 1991 the Navy found to its horror that it faced significant cuts in its budget.

If that weren't distraction enough, the Navy that year suffered from a major self-inflicted blow with the 1991 Tailhook Association scandal.

That was the tense situation in which I found myself upon reporting to Navy headquarters in 1992, where under a new secretary of the Navy, Sean O'Keefe, we finished and released the service's new strategy titled "From the Sea: Preparing the Naval Service for the 21st Century," a white paper that formally enunciated the shift in the Navy's role from a deep-ocean fighting force to one that would operate in the shallow littoral regions of the world.[14] And surprisingly to many observers—given the Navy's historic reputation as the most independent and go-it-alone of the U.S. armed forces—the new strategy paper formally acknowledged the need for improved interservice cooperation and joint operations with the Navy's counterparts in the Army and the Air Force.

By late 1992 the Navy leadership under Admiral Kelso was aware that major changes were mandatory in the way the service planned and organized its combat forces and that there was a serious need to implement reforms that would allow the service to work effectively with the Army, Air Force, and Marines. But there was still little consensus on what these changes meant for the future size and force structure of the Navy, and even less agreement on how we could translate this new policy into the hard specifics of programs and budgets. Finding some agreement and bringing it about was my job.

The Defeat of the Barons

The first task I undertook upon becoming Kelso's resources director in late summer of 1992 was to destroy the enemy within—an archaic and antiquated structure within Navy headquarters that had artificially divided the service into fiefdoms centered on surface ships, submarines, and naval aviation, institutionalizing an internal rivalry among these traditional "warfare communities" at the expense of innovation and combat effectiveness in the overall force.

I assumed that promoting interservice cooperation was a fundamental premise behind the organizational reforms Admiral Kelso expected me to carry out in my new job. I decided early on that one of the most important things I could achieve was to get people to *think differently* about the basic issues with which we had to deal: the size and composition of naval forces, the linkages between weapons and sensors, and the interrelationships between the Navy and the other U.S. armed services.[15] Our first task was to come up with a new scheme to identify the Navy's priorities in making budget decisions as well as to assess combat capabilities. We had to come up with a way to encourage and hasten a new way of thinking about how the Navy would go to war in the twenty-first century.

Thinking Differently About War

For decades Navy planners had worked in a *stovepipe* environment—an organizational structure wherein information and decisions traveled up and down the chain of command of each separate war-fighting community but rarely moved from one domain to another. Traditionally, submariners had planned tactics and developed hardware that dealt exclusively with the submarine force. Naval aviators had concentrated on fielding improved combat warplanes on modernized aircraft carrier designs. Surface warriors focused on the weapons and systems that would expand the combat power of cruisers, destroyers, and frigates. *And rarely did these communities ever compare notes or even talk with one another—much less talk with the other U.S. military services.*

So at the heart of our revolution in the Navy was the effort to redesign the way we would talk about improving naval capabilities. In bureaucratic terms (and, alas, this was a bureaucracy we were trying to transform), we devised a "program assessment structure" that redefined the subjects of discussion away from "platforms"—aircraft, surface ships, or submarines—and toward mission areas. These included priority topics we labeled as "joint" since the new Navy strategy anticipated that future combat operations would involve all of the services performing together as a true team. The topics included "joint strike," "joint littoral warfare," "joint surveillance," "joint space and electronic warfare/intelligence," "strategic deterrence," "strategic sealift," "naval presence," and "special (classified) programs." We called it the "Joint Mission Area Assessment" process.

I was particularly pleased to create a new panel of Navy admirals and Marine generals with the innocuous title of the Requirements Resources Review Board (with the 1990s-era acronym R^3B). Unlike earlier groups, this was not a committee charged with reviewing the staff inputs of junior officers. Rather, we used this forum to brainstorm about a new consensus on what the future size, structure, and character of the Navy should be in the future.

As we began determining the future structure of the Navy, our greatest success came in the adoption of a new planning process that embraced a "joint" mission perspective. We would no longer consider a program or weapons system only in light of the Navy's needs and capabilities, but would take into consideration the other armed services' needs and capabilities as well. For example, advocates pressing for full funding of a given program such as the New Attack Submarine had to demonstrate to all Navy leaders—not just to the submariners—how this proposal strengthened the *joint mission areas of* littoral warfare, joint strike, and other categories. In order to "win" funding, then, a submarine community expert not only had to demonstrate the objective value of his program, but to prevail he also had to learn enough about the other competing Navy budget requests to demonstrate why his was the best candidate.

During this pivotal time, our hardest task was to reach a consensus within the Navy leadership on how to "recapitalize" the fleet in the mid-1990s to maintain current combat capability, shed excess ships and equipment, and in doing so husband sufficient funds that could be used to finance the long-term technological research we deemed essential to modernize the service for the twenty-first century. After intense debate we agreed to shrink the fleet from its Cold War high of just under 600 ships to 340 by the end of 1996, retiring as many vessels as quickly as possible to realize savings from the operating and manpower costs. And we did it disproportionately within the Navy, making a much greater cut in the number of submarines and surface ships and only a very small reduction in the Marine Corps budget and *no* reduction in the number of aircraft carriers.

In return, we were able to save the technological base of the submarine service by financing procurement of the New Attack Submarine (now called the *Virginia*-class nuclear attack submarine). We were able to expand production of the modern, capable *Arleigh Burke*-class guided-missile destroyer from one to three ships per year. We were able to save the aircraft carrier base of twelve flattops. We managed to direct procurement funds into a new class of amphibious ships for our Marine Corps teammates. And we were able to preserve the overall manpower and strength of the Marine Corps itself.

Between the summer of 1992 and early January 1993, we succeeded in reinventing the Navy's long-term plans not only to move from a "blue water" force to a Navy–Marine Corps team prepared to wage war in the littoral regions of the world, but also to take major steps to shed the service of surplus bases, ships, and aircraft. In just five months, between September 1992 and January 1993, we also started the Navy on a historic transformation to meet the difficult challenges of a new, volatile era of U.S. history. For the time being, we were prepared to attain a clear-cut goal: a smaller, technically advanced Navy–Marine Corps force, with a reservoir of budget savings available for future modernization, and a joint force restructured to meet the requirements of joint military operations in the littoral regions of the world, in crises ranging from humanitarian disasters to full-scale war.

On a personal basis, I discovered that there is a very fine line separating an officer's sense of loyalty to his own service that enhances military effectiveness through personal dedication and commitment on the one hand, and the narrow-minded parochialism that erodes overall military effectiveness, on the other hand. I learned how men of quality, dedication, honesty, and commitment can end up on different sides despite decades of friendship, respect, and shared experiences. Because I was pressing the Navy to make radical changes, it was inevitable that some of my long friendships and professional partnerships would be strained, even broken.

The most painful experience for me was seeing the impact that my work as Navy resources director had on my long friendship with Admiral Mike Boorda. Mike and I had been executive assistants together in the Pentagon for many years. We would joke about who was the "world's greatest executive assistant" and had a wonderful, close relationship. When one or the other of us would go on a tour of duty over the years, we'd always try to have a beer together and part with the same phrase, "I'll see you in the next port, shipmate."

But as we rose into the Navy's senior ranks, it became clear that we held radically different professional views. Mike's self-avowed priority was to preserve and protect the size, budget, and structure of the U.S. Navy, his Navy—irrespective of any other consideration—because he deeply believed that the Navy was the core of America's military capability. My view over the years had shifted toward the conviction that we in the Navy needed to implement major changes in order to become more joint—to work better and more closely with the other services. So despite our genuine friendship and affection for one another, Mike and I moved professionally toward the opposite ends of the debate, almost always emerging on different sides of any discussion that had to do with the Navy's force structure or operations. This gap between us widened as we became members of the Joint Chiefs of Staff—Mike as chief of naval operations and I as vice chairman. I can remember the feeling of competition, the sense of ego, and the desire to get "my way" over my old friend. And I knew he, too, felt the same rages about me. I suspect I didn't make his life any easier in my attempts to promote my causes, and I clearly recall the upset I felt as he guarded his interests, which I thought were counterproductive.

I'm glad that our friendship was able to survive that period of intense rivalry and disagreement. I remember with affection and pride Mike's short remarks when I retired from active duty in February 1996. Mike described our relationship without rancor and with honesty. He ended his remarks with, "And I'll see you in the next port, shipmate."[16]

It had been eighteen months since I had regretfully stepped down from a seagoing job to return to the pressure cooker of Pentagon life when in late 1993, Admiral Kelso summoned me to his spacious office along the outer face of the Pentagon E-ring. With a broad smile, the chief of naval operations informed me that my service as Navy resources director was complete and that he was nominating me for a fourth star and a new job as commander in chief of the U.S. Pacific Fleet at Pearl Harbor, Hawaii. I was happy beyond description. There are many good jobs in the U.S. Navy but this is one of the best: going back to the "home" of the U.S. Navy, the Pacific, commanding two numbered fleets, including the forward-based U.S. 7th Fleet in Yokosuka, Japan, and the 3rd Fleet headquartered in San Diego, and one half of the ships, aircraft, and personnel in the entire Navy. My wife, Monika, and I spent several weeks hastily packing our household goods and flew to Hawaii just before Christmas, expecting to begin a major new chapter in our lives together in the storied and historic naval community on Oahu.

We were in Hawaii less than two weeks, unpacking and preparing for the formal change-of-command ceremony, when the telephone rang one morning. It was an Army officer I barely knew, whose faint Polish accent betrayed his background as a childhood immigrant to the United States, a man who had repaid his adopted country with long and distinguished military service.

"Hello, Bill," said General John Shalikashvili, the chairman of the Joint Chiefs of Staff. "I have got great news for you. Your plans have changed. You have been nominated to be the next vice chairman of the Joint Chiefs of Staff." The general went on to say that Secretary of Defense William Perry, Undersecretary John Deutch, and he had studied my record in implementing the Navy reorganization and they wanted me back in the Pentagon as the nation's second ranking officer to work with the Pentagon leadership to meet similar challenges in the U.S. military as a whole.

Even more than commanding the Pacific Fleet, this was the professional challenge for which I had been best prepared over the past decade. But I found myself gazing out my window at the lush palm trees gently swaying under a bright blue sky, thinking of the dim Pentagon corridors and the endless meetings and rancorous debates of the last two years. And I recalled the words of the wise man who once said: No good deed goes unpunished.

Launching a Wider Revolution

When I returned to the Pentagon in January 1994 after my promotion to full admiral and my appointment as vice chairman of the Joint Chiefs of Staff, it was exhilarating but also daunting to find myself a member of the senior leadership of the entire U.S. military. During 1988–89 I had worked as a senior military assistant with Frank Carlucci, the Reagan Administration's last defense secretary, and with his successor, Dick Cheney, appointed by President George Bush in March 1989, so I had considerable appreciation for the challenges near the top of the "world's largest and most complex business." Secretary Perry and General Shalikashvili made it clear to me from the outset that they wanted me to work toward a transformation of the entire U.S. military in much the same vein that I had for the Navy Department. I had no illusions that every step, every recommendation, and every new idea would gore someone's ox, or tromp on some powerful set of toes.

For the next two years I worked hard to focus current and future military budgets on the elements of the Revolution in Military Affairs: the capabilities for improving *battlespace awareness; command, control, communications, computers, and intelligence;* and *precision force.* The bureaucratic terrain on which this campaign would be waged was much larger than that of the Navy alone, and the number of obstacles and opponents would also be much larger. But I was not without resources.

It did not take long for me to gather the essential tools to begin this effort. First, under the Goldwater-Nichols Act and subsequent amendments, the office of the vice chairman had been formally instituted within the Pentagon command organization, and under its two previous occupants the office had matured and strengthened, with the vice chairman becoming a key member of the leadership. I had full legal power to carry out the policy and management decisions we determined. Second, Perry, Deutch, and Shalikashvili were very supportive of my attempts to impose a new system of program and budget assessments aimed at making *joint* military operations, interservice planning, and the spectrum of information technology the top priorities for the Pentagon. Andrew Marshall at the Pentagon Office of Net Assessment, who had long decried the fragmentation and paralysis among the four combat services, remained a powerful ally and strong supporter. Third, a long-overlooked entity used by previous vice chairmen gave me the perfect platform on which to begin my second tour as a military revolutionary: the Joint Requirements Oversight Council (JROC).

First organized as the Joint Requirements and Management Board in 1984, the panel initially was employed to monitor "big ticket" procurement projects and to validate their need. But after the enactment of the Goldwater-Nichols Defense Reorganization Act in 1986, and the subsequent centralization of decision-making authority in the office of the chairman of the Joint Chiefs, the oversight board suddenly had the *potential* to be the launching pad for meaningful defense reforms—and for the American Revolution in Military Affairs itself.

I began my work as vice chairman by implementing a program based on a simple but valuable "lesson learned" from my tenure as Navy resources director: If you want a massive institution like the U.S. military to change, your essential first step is to change people's minds. And the only way to change the minds of a group of seasoned military commanders who have lived their entire careers to the sound of the Cold War drums is to put them in a place where they themselves become an active part of discussion, debate, and thought. And you accomplish *that* goal by literally taking them out of their offices and day-to-day routines for as long as required. We set aside a conference room inside the National Military Command Center in the Pentagon and after a few months of tentative participation, the service "four-star vice chiefs" of staff were actively participating in this new role.

We accomplished some serious business. I had brought the template of joint warfare requirements with me to my new post, and we used it to compile a no-nonsense roster of military priorities for the Pentagon as we looked toward the next ten years and our best analysis of the world situation that would face us then.

Our newest tool was the *Joint Warfare Capabilities Assessment* matrix, which—like the planning model I had introduced to the Navy staff—changed the normal flow of information and program assessment 90 degrees from the traditional way. Whereas the Army had studied and assessed programs and systems related to ground-force combat and ignored naval and airpower themes, and whereas the Navy had ignored the needs of land and air services while focusing on maritime issues, our deliberations examined the gamut of military power subjects from the perspective of interservice (joint) commonality. We also addressed the most likely serious threats to the United States we anticipated over the next two decades. The Joint Warfare Capabilities Assessments included

- joint strike
- land and littoral operations
- strategic mobility and sustainability
- sea, air, and space superiority
- deter/counter proliferation of weapons of mass destruction
- command and control
- information warfare
- intelligence, surveillance, and reconnaissance
- regional engagement/presence
- joint readiness

Through the process of issue identification and debate, it was our goal that the matrix of these Joint Warfare Readiness Capabilities would connect the armed services' perception of emerging military *needs* with the Defense Department's decision-making sequence for *meeting* those needs through new or reemphasized budget support and funding priorities. A clear vision and carefully agreed upon priorities were essential since these decisions were being made at a time of unprecedented military budget cuts.

The second tool was a relatively new milestone in the annual budget process called the *Chairman's Program Assessment.* As part of the decision-making and planning effort, the chairman of the Joint Chiefs and his staff, supported by the JROC, formally recommended priorities and specifics of both spending cuts and increased funding directly to the secretary of defense. I was pleased that throughout my two-year tenure as vice chairman, spending on programs that will make up the "system of systems" leading to a Revolution in Military Affairs steadily grew in reflection of our recommendations.

Our central premise was that in an era of shrinking force structure and diminishing "top-line" defense budgets, the most effective response to an emerging security threat would come from leveraging the synergy of disparate military programs and components that had never been connected in the previous era of *military service unilateralism.* To a much greater extent the senior military four-star leaders were all in this together. Although this new cooperation among the military service chiefs lessened the influence of the various civilian staff offices at the office of the secretary of defense and the civilian secretariats of the Air Force, Army, and Navy Departments, I remain convinced that the reward for our efforts was a planning and budgeting process that truly aimed at improving the strength of the U.S. armed forces in an era of limited financial resources.

We made several headlines. One of the most controversial agreements we hammered out as a result of our extended discussions and field trips to visit the regional war-fighting commands, was the decision to halt production of the B-2A stealth bomber at 20 aircraft (later expanded by President Clinton in the 1996 election campaign to 21). I had always thought the B-2 could play a pivotal role in the delivery of inexpensive (and nonnuclear) precision-guided bombs. However, we concluded that with the steady decline in defense spending, allocating the additional billions of dollars for the additional B-2As instead of for other, equally urgent procurement programs would be devastating to our overall plan. (The Air Force, which for decades had touted bomber strength above all other aircraft, was in full agreement with this decision. Its top priority was to secure steady funding for planned advanced tactical aircraft programs.)

We also made several decisions to set in stage major programs that will not emerge from the research-and-development process for several years, but that, if successful, will form pillars of the Revolution in Military Affairs. These include an array of individual programs aimed at establishing Unmanned Aerial Vehicles programs, providing theater ballistic missile defense, and enabling the ongoing airborne laser project to fit a high-energy laser on a modified Boeing 747-400 freighter aircraft to shoot down ballistic missiles during their initial boost phase.

Occasionally we came up with an idea that convinced me we could attain major breakthroughs in the management and use of military force. The example I cite most regularly stemmed from an ad hoc move the Navy made in the 1980s during the Iran-Iraq War when we needed to position some special operations forces in the Persian Gulf. When friendly nations (out of fear of angering Iran) refused to grant us permission to bring our forces in, we obtained and operated a large mobile platform anchored offshore that consisted of several large civilian barges lashed together. The setup was primitive and ad hoc, but the mobile base served a useful purpose as we played a low-level game of cat and mouse with Iranians attacking neutral oil tanker shipping.

This concept came back to me not long after I became vice chairman and I began to think: Why not develop a full-scale, mobile offshore base? The shipbuilding capability has long existed to construct such mobile platforms; large mobile oil-drilling platforms validate that the basic concept works. And the cost of each platform would be less than that of an aircraft carrier. After a quick staff study I proposed for consideration the concept of a "mobile offshore base" design consisting of three or more 600-by-600-foot sections that could maneuver with internal propulsion at speeds up to 6 knots. The sections would readily join together to form a very large floating base with a normal full-length (4,000–10,000 feet) runway on top, space for massive aircraft and machinery repair facilities, logistics storage, and troop berthing spaces and room for other military units and gear. The mobile offshore base would be able to accommodate Navy and Air Force combat aircraft, several thousand Army or Marine troops, and a wide number of medium-sized transport and supply planes much too large to land on aircraft carriers.[17] And instead of operating in a region for only three months at a time (like the carrier fleet), the mobile bases would operate in an area as long as we needed them there. I envisioned three of these bases permanently assigned to strategically vital regions such as the Western Pacific, the Persian Gulf, and the Mediterranean Sea. Alas, the Navy and Air Force as one rose up in indignation. The carrier admirals were against the mobile offshore base because they saw it as a threat to the aircraft carrier force (although it was not), and the Air Force generals saw the mobile offshore base as a threat to their units as well. The concept is technically still alive today, but it languishes in a remote corner of the Pentagon research community.

And some victories were only moral victories. After hundreds of hours of meetings and debates, I persuaded the JROC to formally recommend a minimum of $60 billion per year as the goal for procurement of weapons and equipment to create the "system of systems" that would embody the Revolution in Military Affairs, and for the important modernization programs required. We made that requirement a very large stake and we pounded it into the ground. Alas, since

1994–95 the record of the White House and Congress in supporting that important investment has been characterized by failure and procrastination. One recent analysis concluded that the military is heading for a "train wreck" by the year 2010 largely because procurement funds have continuously fallen $20 billion per year below our goal of $60 billion.[18]

A Personal Exit

On February 27, 1996, I put on my Navy dress uniform for the last time and mounted a stage at the U.S. Naval Academy, where I had first begun my naval career as a "plebe" midshipman in the summer of 1958. After two years as vice chairman, I believed I had made some progress on the mission that Secretary Perry, Undersecretary Deutch (later succeeded by John White), and General Shalikashvili had outlined for me, but I had chosen to decline an offer for a second two-year term as vice chairman. A new life beckoned in California as corporate executive of a scientific and research firm that was deeply involved in many issues that are vital to the future of the U.S. military. I turned over my office to Air Force General Joe Ralston, an outstanding officer and military commander, confident that he would try to continue the work I and my colleagues had started on the Revolution in Military Affairs.

I left active duty convinced that the U.S. armed forces had made the first important steps toward a major transformation that would ensure our combat superiority well into the twenty-first century. We had created an analysis and decision-making forum that permitted all of us to shed our service parochialism and make the hard decisions on the basis of optimum need for the U.S. military as a whole. The only question was, would the revolution continue and succeed? At that moment, I was willing to let it be someone else's priority.

CHAPTER 5

THE LESSONS OF KOSOVO

For seventy-eight days and nights in the spring of 1999, after diplomats had failed to halt Serbian oppression of the ethnic Albanian majority in Kosovo, the United States and its NATO allies conducted a massive air war against Yugoslavia. Finally, on June 10, Serbian dictator Slobodan Milosevic agreed to NATO's demands that he withdraw his military and special police from Kosovo and allow a massive multinational peacekeeping force to enter the region. No sooner had his surrender been announced to the press than the debate began in earnest over what kind of victory the Clinton Administration and the NATO alliance had won in Kosovo—and, indeed, over whether it had been a victory at all.

Behind the scenes, U.S. military experts were deeply disappointed in the way the Kosovo operation had played out. "The happy talk about successes up front is just untrue," one of my former colleagues in Europe told me. "This was nothing more than siege warfare on the model of the 16th century, with the use of airpower as artillery because the other critical elements of power—ground forces—were removed from [NATO's] arsenal."[1]

It did not take long for the initial euphoria over the Serbs' agreement to withdraw from Kosovo to evaporate.

- General Wesley Clark, the brilliant U.S. Army general who led the NATO coalition against Serbia, was publicly humiliated by the

Pentagon when officials in the office of Secretary of Defense William Cohen revealed that Clark would leave his post three months early so that my successor as vice chairman, General Joseph W. Ralston, could replace him as NATO commander. (Many think this apparent release of the winning commander was badly mishandled.

- Many questioned whether the NATO action may actually have hardened Milosevic's resolve to hasten the ethnic cleansing—and thereby led to the de facto needless deaths of tens of thousands of Kosovars and the suffering of hundreds of thousands of others who were forced from their homes.
- Serious questions arose concerning the coalition's inability to *truly* integrate forces from all services, even within the U.S. forces.
- The world's most professionally executed strategic air campaign was unable to discriminate between real ground vehicles (tanks, armored vehicles, etc.) and mock vehicles (plywood, dead hulks, etc.) or to quickly and completely knock out the Serbian Integrated Air Defense System.
- The return of hundreds of thousands of Kosovar civilians to their homeland under NATO protection, after having been ejected by Serbian soldiers and special paramilitary police, was marred by a spate of violence against the minority Serbian population in Kosovo. This led to a second (albeit smaller) refugee crisis as thousands of Serbs fled and those who remained endured brutal harassment by the returning Kosovars.
- Relations between the United States and both China and Russia became strained over the accidental bombing of the Chinese Embassy in Belgrade and over Russia's strong opposition to the NATO air war.
- Many questioned whether the United States and NATO had failed to achieve two of President Clinton's stated goals: "to deter an even bloodier offensive against innocent civilians in Kosovo and, if necessary, to seriously damage the Serbian military's capacity to harm the people of Kosovo."

Some skeptics have warned that it may be a mistake to draw too many conclusions from Kosovo. "One must resist the temptation to read too much into a conflict that pitted half the world's industrial might against a nation with a gross domestic product less than Idaho's," defense analyst Michael G. Vickers wrote in *USA Today* shortly after the end of the bombing.[2] But as Vickers and many other commentators quickly realized, Kosovo demonstrated the twin paradigms of conflict in the new century.

First, the profusion of civil wars and internal conflicts throughout the world poses a difficult political challenge to the United States and its allies that no degree of military technology alone can resolve. Second, the technology itself—as it matures and enters the active force in the form of weapons and sensor systems—will continue to put pressure on U.S. military leaders to adjust doctrine, tactics, and even basic military organizational structure to maximize the technology's effectiveness. For this reason, I believe Kosovo will stand as a symbol of the unavoidable, difficult decisions that face us in the years to come.

I also believe that it is important for us as a nation to take a close look at what happened in the conflict—what *really* happened. The air war over Serbia in 1999 showed that there are real weaknesses in our military and that these weaknesses will put our nation in jeopardy unless the U.S. armed forces embrace the Revolution in Military Affairs and the joint forces culture that goes with it.

While the operation did not employ the full range of American ground and naval power, the political stakes for NATO were immense. The aerial campaign was the largest use of force by the United States and its allies since the Persian Gulf War. In the aftermath, Air Force and Navy aviation officials conceded that their units and personnel had faced the stresses of a conflict that—for them—constituted a major theater war, cost billions of dollars, and left them facing several years of recovery to replenish the smart weapons stockpiles, and to return forces and equipment to their prewar states of combat readiness.

Coming just a year before a U.S. presidential election campaign and at a time of renewed congressional attention to the "declining state of our combat readiness," the Kosovo engagement illuminated problems with the U.S. armed forces that our current leaders have been unable or unwilling to address.

From my own experience as a former naval officer and senior U.S. military commander, I believe U.S. political leaders, military commanders, and concerned citizens face one critical task in assessing the lessons of Kosovo: to be rigorously honest, open-minded, and objective in defining the actual strengths and weaknesses in the way our military fought the war and in the way our government and federal bureaucracy handled it at home.

While this book is an overview of the current status of the U.S. military and its pressing need to fully implement the Revolution in Military Affairs, a key lesson from Kosovo (like Beirut, Somalia, and Vietnam before) has nothing to do with technology or information: Neither the technology revolution nor the dedication and professionalism of our men and women in uniform can overcome the consequences of poor political decisions or poorly executed policies.

Objective observers looking back on the Kosovo campaign can readily see through the short-term political "spin" that described a unique NATO airpower victory that reversed Serbian aggression in Kosovo. The facts dictate a harsher conclusion.

Victory presumes that the victors secured the political objectives that led them to launch the military campaign in the first place. A dispassionate review of the facts shows that the U.S. and allied forces failed to achieve most of President Clinton's and NATO's avowed objectives in the Kosovo conflict.

I also challenge those who say the air campaign heralded an era in which airpower alone is sufficient to prevail against an aggressor nation. The air campaign was professionally drawn up and carried out. Over half of the Serb military's petroleum supplies were eliminated and essentially all of the country's refining capability was destroyed; electrical power in Belgrade was seriously degraded. However, evidence from the aftermath of the operation revealed that NATO bombing was *ineffective against the Serb campaign of terror and ethnic cleansing in Kosovo* until late in the two-and-a-half-month conflict, and the bombing scored truly meaningful results against the Serbs only after a reconstituted Kosovo Liberation Army of fewer than 20,000 began a concerted series of ground attacks that flushed Serbian units from their hiding places. The coalition's air operations succeeded only in the context of the old paradigm of industrial-age warfare—as a force that slowly ground down Serbian military capability through attacks on fixed military sites and civilian facilities, accompanied by "collateral damage" incidents in which innocent Serbs and Kosovars (and Chinese diplomats) were bombed instead of the actual military targets. The Kosovo conflict showed that airpower remains most effective when used as one component of a multifaceted military operation, not as the only element.

Even Milosevic's ultimate decision to end the war of attrition against his country by ordering Serbian military and special police out of Kosovo failed to constitute a NATO victory, since the alliance's initial objective was to preserve a stable, pluralistic population of Serbian and Albanian communities within the province—not to replace ethnic cleansing of Kosovars with ethnic cleansing of Serbs. The outcome of the air war was the destruction of the Kosovo we wanted to safeguard and the creation of a political vacuum that will require the presence of U.N., U.S., NATO, and Russian peacekeepers in Kosovo for years to come. And what are the fruits of our "victory"? Renewed and hardened political tensions between the United States and Russia; a major setback in the ongoing political entente with China; and yet another open-ended deployment of American and Western European peacekeepers to a region still held hostage by ethnic enmities. That doesn't sound like a victory to me.

OUR KOSOVO OBJECTIVES

Let's briefly review the objectives of the United States and its allies with regard to Kosovo in early 1999.

Having tentatively brokered a cease-fire in the Serb campaign against the Kosovo Liberation Army in late 1998 (and dispatched unarmed "observers" to the troubled province), the United States and our European partners in early 1999 were concerned over the rebuilding of the KLA guerrilla force that winter and a sudden upsurge in attacks and reprisals on both sides. As early as February 2, 1999, senior U.S. officials warned President Clinton and Secretary of Defense Cohen that the United States and NATO would have to step in directly. Director of Central Intelligence George J. Tenet told a congressional committee that Serb-Albanian hostilities would probably get much worse unless the "international community succeed[ed] in imposing a political settlement" and NATO deployed ground forces "to enforce its implementation and deter new fighting."[3]

Given the deep hatreds that existed within Kosovo on both sides, it was nearly futile to try to find—much less dictate by diplomatic fiat—any common ground between the Serbs and the ethnic Albanians. Clearly the Albanians wanted outright independence. And the Serbs wanted the Milosevic regime to do whatever it took to preserve their security and political control from an ethnic Albanian takeover.

In the weeks leading up to Operation Allied Force, the United States watched from the sidelines as France and Great Britain sought a settlement in a peace conference at Rambouillet, France. But when the talks bogged down (to no one's surprise), President Clinton and Secretary of State Madeleine Albright delivered a diktat to both sides: agree to a NATO peacekeeping force of 28,000 troops that would occupy Kosovo. At that time, neither the United States nor NATO formally supported the 10,000-man Kosovo Liberation Army, and both opposed the Albanian force's campaign for full independence from Serbia. As a report in *The New York Times* on the eve of the Rambouillet conference succinctly noted, the KLA was viewed as something less than a potential ally:

> American and NATO officials are already working to reduce the flow of arms to the guerrillas from Albania and are using sophisticated surveillance techniques along the mountainous border. While no one believes that border can be sealed, the officials have talked with the government of Albania about a more intensive international effort to scrutinize the Tirana airport and the Albanian ports of Vlora and Durres to prevent the smuggling of arms and money to the ethnic Albanian forces in Kosovo.
>
> Kosovo Liberation Army bank accounts can be closed, the officials said, and communications and logistics disrupted. The ethnic Albanians have also been told that the current international support and sympathy for them and their fight for self-rule could quickly disappear.[4]

The U.S. government's objectives for Kosovo, then, were clear: (1) to protect all ethnic groups from violence and oppression; (2) to maintain Kosovo as an integral province of Yugoslavia; (3) to prevent the internal political struggle from spilling over into adjoining countries such as Albania, Macedonia, and Greece, where ethnic tensions could lead to instability or draw those countries directly into the conflict (such as Greece supporting the Kosovo Serbs and Turkey abetting the Muslim Kosovars); and (4) to maintain our post–Cold War entente with Russia despite Moscow's historic ties to the Serbs.

As he talked about the NATO aerial campaign, Clinton told the American people on March 24, "Our strikes have three objectives. First, to demonstrate NATO's opposition to aggression and its support for peace. Second, to deter President Milosevic from continuing and escalating his attacks on helpless civilians by imposing a price for those attacks. And third, if necessary, to damage Serbia's capacity to wage war against Kosovo in the future by seriously diminishing its military capabilities."

Milosevic's refusal at Rambouillet to allow a NATO peacekeeping force into Kosovo—in effect to surrender a province of his nation to outside political and military control—made NATO's military intervention inevitable. And the United States, which had launched a brief but fierce air war against Iraq late in 1998, went to war in the air for the second time in three months.

It is instructive to note how unprepared the Clinton Administration and its European allies were for Serbia's refusal to knuckle under to NATO. On the night the air strikes began, Secretary of State Madeleine Albright, appearing on *The NewsHour with Jim Lehrer,* said, "I don't think of this as a long-term operation. I think that this is something . . . that is achievable within a short period of time." Two weeks after the air campaign began, U.S. State Department officials admitted in one press report that they were surprised by Milosevic's tenacity and by the Serbian leader's immediate call for his people to undertake "ethnic cleansing" of the Kosovars. In the days that followed, hundreds of thousands of Kosovars began fleeing into Macedonia and Albania. It is worthwhile to review in detail one key report published by *The Washington Post*:

> Albright and her closest aides expected Milosevic to behave like a "schoolyard bully," as one senior official put it, backing down after a few punches were thrown. They admit they were unprepared for the scope and speed of the deportation campaign. By contrast,

senior Pentagon officials expressed doubts before the war that Milosevic could be moved by air power, and CIA Director George J. Tenet warned that the Serbs might respond with a campaign of ethnic cleansing.

While administration officials have argued that the bombing was necessary to try to stop a campaign of ethnic cleansing that began a year ago, they also said that nobody predicted that Milosevic would respond by forcing civilians onto trains and deporting them, a scene not witnessed in Europe since the depths of World War II.

"As we contemplated the use of force over the past 14 months, we constructed four different models," one senior official said. "One was that the whiff of gunpowder, just the threat of force, would make [Milosevic] back down. Another was that he needed to take some hit to justify acquiescence. Another was that he was a playground bully who would fight but back off after a punch in the nose. And the fourth was that he would react like Saddam Hussein," the president of Iraq, who hunkered down through Operation Desert Storm in 1991 and still holds power.

"On any given day people would pick one or the other," this official said. "We thought the Saddam Hussein option was always the least likely, but we knew it was out there, and now we're looking at it."[5]

In a televised address on March 24, shortly before the air war began, President Clinton (echoed by other administration officials) emphatically declared that the alliance would not employ ground combat power against Serbia, saying: "If NATO is invited to [send a peacekeeping force], our troops should take part, . . . but I do not intend to put our troops in Kosovo to fight a war." Three weeks later, two staunch Clinton supporters, Senator John Kerry, a Democrat from Massachusetts, and Senator Joseph Biden, a Democrat from Delaware, would harshly criticize Clinton's decision to foreswear ground troops. Kerry called the decision "foolhardy"; Biden, "a mistake."[6]

Three weeks into the air campaign—and with 1 million Kosovars on the run from their homeland—the administration seemed to reverse itself, hinting that a ground war was still under consideration. Milosevic had only to watch the European news service Sky TV to know that the rhetoric was hollow. There were no signs of any preparation for a NATO ground war, whether to occupy Kosovo or to separate the province from Serbia with a *cordon sanitaire*.[7] The absence of ground war plans provoked an eloquent criticism from my friend Richard Haass, director of foreign policy studies at the Brookings Institution and a former National Security Council official with the Bush Administration. Haass wrote:

> Increasingly it seems that the Clinton Administration's foreign policy is intended to minimize risks rather than maximize results. The result is bad politics and bad policy. Take the debate about ground troops in Kosovo. After weeks of ruling out their use, the Clinton Administration is now sending mixed messages.
>
> On Sunday (April 13, 1999), Administration officials, including Gen. Henry Shelton, the Chairman of the Joint Chiefs of Staff, suggested that plans for ground troops exist and could be taken off the shelf at any time. Yesterday, Defense Secretary William Cohen said that the air attacks are increasingly effective, while reiterating General Shelton's comments about ground troops.
>
> What's going on here? It seems as if foreign policy is being driven by public opinion. News photos of suffering Albanian refugees have had an enormous impact on the American people; opinion polls indicate that about half of them now favor sending ground forces into Kosovo. But the Administration also seems to have no confidence that popular support would survive the first casualties.
>
> This is no way to make foreign policy—or win a war. It is one thing to rule out ground troops because they are not needed. It is something else again to reject them out of fear that the American people will not back their use.
>
> Indeed, history suggests that Americans will support using ground troops, even after the country suffers casualties. The public even supported the Vietnam War for more than a decade despite the horrible costs.[8]

Even after two months of air strikes, Defense Secretary Cohen still warned that any move toward a ground war could split NATO and undercut allied support for the air operation. He joined other Clinton Administration officials in refusing even to talk about it. "There is no consensus within the alliance for a ground force," Cohen said on CBS's *Face the Nation* on May 16, and press reports confirmed that NATO had done little to assemble a ground force.[9]

Clinton repeated his fear of a ground war in Kosovo on March 31, when in an interview he said, "The thing that bothers me about introducing ground troops into Kosovo and into the Balkans, is the prospect of never being able to get them out."[10]

By early April senior U.S. military commanders were admitting the ugly little secret about Kosovo: there was no plan either in the White House or in the Pentagon for a ground war. Had we been sincere about opposing Milosevic, the President would have taken a leadership role—winning American public support and persuading NATO leaders to accept what was necessary if we were to achieve our objectives, namely, two or three days of crippling air strikes, coupled with a credible ground force—deploying combat units, marshalling logistics, and constructing advance bases along the periphery of Kosovo. This did not happen. Milosevic and his generals knew it, and a million Kosovars suffered the consequences. With no threat of a NATO ground force, the Serbs were free to disperse their units throughout Kosovo to avoid NATO air strikes, while they killed and tortured ethnic Albanians. One expert summed up the situation clearly: "You need the early decision, to get troops trained and ready," said retired Army General George Joulwan, a former NATO commander. "This should have been started in October (1998)." Instead, six months later, a decision on ground troops still had not been made, and Serbian troops had shrugged off thousands of NATO air strikes and expelled countless more ethnic Albanians.[11]

AN AIR CAMPAIGN BY DEFAULT

The war the United States and NATO did fight in Kosovo—Operation Allied Force—was a war by default: an air-only war designed more to avoid U.S. and allied casualties than to achieve our leaders' stated objectives. This approach not only yielded the strategic and tactical initiative to Milosevic, but also undermined the moral argument NATO had used to justify going to war in the first place. Writing in the *U.S. Naval Institute Proceedings*, Dr. Michael Evans noted, "This was supposed to be a war of human values rather than political interests, but by ruling out ground forces from the outset, NATO signaled to Belgrade that the Kosovar people were not worth the life of a single NATO soldier."[12]

Superior technology, sound organization, and well-trained personnel are essential if a modern military force is to win a war or hold its own. But even the best military force can be squandered if it is sent on a mission that simply is not suited for resolution by military means. Likewise, cowardice and lack of foresight in our political leaders' decisions over how to use military force can lose a war that should have been won.

As we seek to bring about the Revolution in Military Affairs, then, we have to keep in mind that it will not be enough to remake our military technologically or to forge true cooperation among our four armed services. We as a nation will also need to reassess how we make strategic decisions within our alliances and how we decide whether and in what manner to use military force in crises not directly involving American citizens or territory.

Operation Allied Force brought to light some serious flaws in the U.S.–NATO structure, and these have led to the long stalemate that has followed the Serbian pullout from Kosovo.

War by Consensus

The NATO alliance in its fifty-year history has been successful at forging a political consensus among members such as Greece and Turkey, who historically have distrusted one another. This consensus has been led by the U.S. government. But when it came time to oppose Serbia, the management-by-committee structure in place in Brussels made going to war impossible.

It was the inability of the alliance members to agree on the use of ground forces—and not only the White House's fear of U.S. public opinion—that led the alliance members to fight an air war by default. As a perceptive *New York Times* analysis pointed out, Clinton Administration officials had shied away from exerting leadership in NATO over the question of the need for ground forces in Kosovo because they feared a replay of the bloody firefight in Somalia in 1993 in which 18 U.S. soldiers and airmen died:

> "The Administration lost faith in the usefulness of ground troops in 1993, after 18 soldiers were killed in a failed raid in Mogadishu," said Ivo Daalder, who was a Bosnia specialist in the White House in the mid-90's and is now at the Brookings Institution, a research group in Washington. "They believe that Somalia demonstrates conclusively that you cannot have any casualties," Daalder said. "They take this as a matter of faith."
>
> In running the [1995] Bosnia combat, General (George) Joulwan recalled, he constantly had to fight Clinton's assumption that a majority of the American public demanded a zero-casualty war. "The problem with the White House is that rather than say why did casualties occur, they just said we can't take casualties any more," General Joulwan said.[13]

During Operation Desert Storm, the U.S.-led coalition gave General H. Norman Schwarzkopf wide latitude to choose the strategy and tactics the coalition would use to eject the Iraqi Army from Kuwait. While President Bush (working through Secretary of Defense Dick Cheney and Joint Chiefs Chairman General Colin Powell) worked with our allies to define Desert Storm's objectives, Schwarzkopf made the key military decision: the kind of force the war required, even if it meant doubling the numbers of ground and naval units deployed to the Persian Gulf region. In Kosovo, a sad contrast, General Wesley Clark had to act with the bare minimum of what the NATO ministers would allow him. And it is arguable that without U.S. leadership, there was no alternative to building what some call a "two-hump camel." That is no way to manage an alliance, and it is no way to mount a war.

As a result, the NATO Military Committee in Brussels—although motivated by genuine concerns to minimize loss of life among both Serb civilians and NATO pilots—had no choice but to impose impossibly strict rules of engagement. Brussels forced the aviators to fly at 15,000 feet without adequate sensors to identify targets on the ground; they took the task of selecting targets away from NATO military leaders and put it into the hands of NATO senior policymakers! One Pentagon official said anonymously that these rules of engagement undid the decision-making structure in the U.S. armed forces and NATO. "We're in an alliance, though," he told *Air Force* magazine. "So this is how we have to do it."[14] He failed to add that the United States, as the historic leader of the NATO alliance, had the wherewithal to insist on a military campaign based on realistic military premises. In fact, we ceded leadership to others, notably British Prime Minister Tony Blair, who in the past would have avoided a military posture that might have undermined transatlantic statesmanship.

General Richard Hawley, a highly decorated and respected Air Force general who was in charge of Air Combat Command during the Kosovo conflict, publicly lamented the straitjacket imposed on U.S. and NATO pilots during the air campaign. "Air power works best when it is used decisively," he told reporters on April 29, 1999. "Shock, mass are the way to achieve early results. Clearly because of the constraints in this operation, we . . . haven't seen that (effect) at this point."[15]

A U.S.–NATO Military Gap

Kosovo confirmed a significant gap between the military capabilities of the United States and its European allies that could soon lead to serious friction over how to share defense burdens. Even as NATO headquarters issued claims of victory, behind the scenes U.S. officials were realizing that Western Europe has fallen seriously behind the United States in the use of precision-guided weapons, satellite reconnaissance communications, and other modern technology—to the point that the European forces are finding it more and more difficult to fight the way we can. This portends serious problems in any future conflict involving NATO, since a root premise of coalition warfare is that the partners be able to work together and that their military components—whether fighter squadrons, ground units, or naval vessels—be coordinated seamlessly.

"The Kosovo war was mainly an experience of Europe's own insufficiency and weakness," German Foreign Minister Joschka Fisher told the *Washington Post.* "We as Europeans never could have coped with the Balkan wars . . . without the help of the United States. The sad truth is that Kosovo showed Europe is still not able to solve its own problems. We have to accept the consequences and hope that Europe can grow from this crisis."[16]

The *Post* analysis noted that the United States now devotes $35 billion a year to creating the kinds of advanced weapons and intelligence-gathering systems used over Yugoslavia—a figure that U.S. military leaders have been saying for years is seriously deficient given our needs to modernize the force—while European alliance members have spent even less, $10 billion a year—a sum spread out over dozens of different national projects.

Perhaps the lessons from Kosovo will be the prompting the United States and NATO need to modernize their forces. One positive development came last spring when members of NATO's Conference of National Armaments Directors showed support for developing an airborne ground-surveillance system such as the Joint Surveillance and Target Radar Systems (JSTARS) which would be owned and operated by the alliance. Under the proposal, five countries, led by the United States and Norway, would consider how NATO can integrate an existing JSTARS airborne radar for NATO's use, *The Wall Street Journal* reported last spring. The proposal had been bogged down in European politics for years. "The Pentagon has warned NATO that defense cuts have stretched the U.S. so thin that it would be difficult to deploy its arms quickly should a crisis erupt," the report added.[17] The bad news is that the eight-year debate over a NATO ground-warning system is still stuck, even though Kosovo painfully underscored the need for such a system. (The NATO technology lag is not an exception. Similar problems have festered for years in the our alliances with South Korea and in the Middle East.)

While NATO focused on Kosovo, the United States had no such luxury: as it was rushing aircraft, surveillance systems, and other combat units to southeast Europe, General Wesley Clark had to struggle to get military assets from other regional U.S. military commanders. The number of U.S. and NATO aircraft committed to Kosovo peaked at 1,045 on June 3, 1999, more than twice the number of aircraft that launched air strikes on March 24. The U.S. Air Force, Navy, and Marine Corps contributed 720 aircraft, and the rest of the alliance came up with

325 aircraft. The beg, borrow, and steal effort it took to assemble that number of aircraft, and, more important, the weapons and surveillance systems to support the air campaign, made it clearer than ever that the current U.S. military is stretched very, very thin. The effort also showed (as I have pointed out) that our military is still stuck in the Cold War paradigm of giving top priority to the high-visibility weapons "platforms" (aircraft, ships, major ground combat vehicles) while skimping on precision-guided ammunition, surveillance systems, sensors, and command-and-control systems. A review of Kosovo events is a roster of serious deficiencies in current U.S. military capability:

- *Precision-guided munitions.* The Air Force used two of its key "smart" weapons in Kosovo, the satellite-guided Joint Direct Attack Munitions (JDAM) and the Conventional Air-Launched Cruise Missile (CALCM), which uses Global Positioning System navigational satellites to carry out pinpoint attacks in all weather conditions. In both cases the air service—ordered to avoid unnecessary risk to pilots and to minimize incidents of collateral damage to civilian targets—relied heavily on these precise, stand-off weapons as the air campaign continued. And within weeks, the service said it was running out of its prewar supply of the weapons: It had fired a majority of its inventory of 900 Joint Direct Attack Munitions dropped from B-2 stealth bombers, and had less than 80 of the 3,000-pound cruise missiles that are launched off B-52 bombers remaining.
- *Specialized combat aircraft.* The U.S. Air Force, Navy, and Marine Corps provided the vast majority of high-technology combat aircraft used in Kosovo, and the stress of the deployment on aircraft and personnel was severe. These aircraft systems—which manifest the current technological lead of the U.S. military—included the E-8 JSTARS ground radar plane, the E-3 AWACS aircraft, EA-6B Prowler electronic jamming aircraft, and the U-2R reconnaissance aircraft. In all of those categories, officials reported that the relatively small number of planes meant that both the aircraft and crew personnel were being heavily overworked as they were sent into the campaign. The JSTARS training program was all but halted because all of the senior instructors were detached for temporary duty in Kosovo, and the EA-6B community waived the time

limits on back-to-back deployments as NATO urgently called for additional aircraft and crews. General Hawley, the air combat commander responsible for Air Force aircraft including the B-2, F-117, F-15, and F-16, sadly pronounced the legacy of Operation Allied Force last spring when he warned that the entire strike arm of the service would need a lengthy "retrenchment" to rejuvenate the Air Force following its overextended deployment to Kosovo.[18]

- *Other operations forestalled.* General Wesley Clark, the NATO commander, was forced to withdraw so many combat aircraft from Turkey to be used in the Balkans that the ongoing U.S.-led Operation Northern Watch patrols over Iraq—a U.N.-mandated operation since 1991—all but ceased for the first two months of the Kosovo conflict. The Navy was forced to remove every last aircraft carrier from the western Pacific, including the tense Korean peninsula; and the USS *Theodore Roosevelt,* when it arrived in the Adriatic last spring, was operating with 400 fewer sailors than its complement requires. An Air Force munition ship normally based at Diego Garcia in the Indian Ocean, which supplies aircraft bound for the Persian Gulf or the western Pacific, was diverted to Europe to supply the Kosovo air strikes.
- *Reservists required.* The Pentagon's fleet of KC-135 and KC-10 refueling planes was spread so thin by demands from Kosovo that in late April President Clinton ordered the first of 33,000 reservists called up to sustain the round-the-clock bombing runs. And shortly before the June 9 cease-fire, the Pentagon was forced to impose a controversial and unpopular "stop loss" order preventing pilots and other critical personnel from leaving active duty at the expiration of their commissions or enlistments.[19]

We must keep in mind that the challenges confronting the U.S. military today will become impossible to overcome in fifteen to twenty years since the current inventory of ships, aircraft, and ground vehicles is *not* being replaced and the force will shrink by 25 to 40 percent. The limited successes of new weapons and surveillance and sensor systems in Kosovo fell far short of a Revolution in Military Affairs.

FLAWS IN COMBAT TECHNOLOGY

Kosovo, like every other war, was a "come as you are affair," and the U.S. military was captured in its current flawed state—rigidly structured for the kind of defensive, terrain-holding war against the Soviet Union that ceased to threaten us with the collapse of communism, but unable to mount joint operations and sorely lacking in information-gathering, surveillance, and communications systems. Our military was so ineffectual in the Kosovo conflict that for seventy-eight days Slobodan Milosevic and his light ground forces bedeviled the most powerful military alliance in the history of the world.

I'll offer three examples that underscore this point. The first was our inability to deploy enough aircraft to pierce the clouds and fog that cloaked the Serbian forces in Kosovo. For the first two weeks of the air campaign, NATO was flying only about sixty to eighty strike missions per day as a result of two factors. First, there were no surveillance and reconnaissance systems that could identify ground targets through the thick cloud cover and pass that information in real time to U.S. and allied strike aircraft. Second, the alliance lacked the political determination to mount the necessary number of sorties (individual aircraft missions) on strategically vital targets to overwhelm the Serbs and force Milosevic to reverse his terror campaign in Kosovo. As a result, Milosevic hunkered down and the campaign reached a stalemate.

The second, and equally serious, deficiency stemmed from the structural rigidity of U.S. Army forces in Europe. Apart from a brigade-sized Marine Expeditionary Unit deployed offshore on U.S. Navy ships, the Pentagon had no "light" ground units available for a quick entry into Kosovo, not even the heralded 82nd Airborne Division and other units of the XVIII Airborne Corps—the Army's self-styled "911 force" for world emergencies.

Just two weeks after the Kosovo conflict ended, General Eric Shinseki took command of the Army as its new chief of staff. His first official communiqué to the troops was a candid and refreshing assessment of the structural flaws Kosovo had brought to light. Calling on the ground service to be "more versatile, agile, lethal and survivable" with "early entry forces that can operate jointly, without access to fixed forward bases [but that] still have the power to slug it out and win campaigns decisively," the general described what NATO so badly needed and did not possess when it went to war on March 24. "Our heavy forces are too heavy and our light forces lack staying power," Shinseki said.[20]

The third flaw—and the most serious one—emerged in the sorry saga of Task Force Hawk, the attack helicopter force dispatched to Albania in anticipation of a nose-to-nose NATO fight to expel the Serbs from Kosovo.

Nevertheless, our dedicated military commanders and pilots improvised well. The United States deployed more than 80 percent of the 1,045 combat aircraft used in Operation Allied Force. Meanwhile, the Europeans—coming to the conflict with what assets they possessed—flew combat strikes in missions not requiring high-tech navigation and all-weather pinpoint weapons, provided host-nation support, and deployed the bulk of the 50,000 ground troops in the NATO-led peacekeeping force. This division of roles and responsibilities worked in Kosovo, but there is a need to give considerable thought to the force structure of the alliance to ensure that future operations are better organized. (1) All members of the alliance should recognize that the United States possesses the mass and technology to build an integrated "intelligence umbrella" over any battlefield in the future, a capability to communicate this information to warriors (both American and European) on every battlefield, and the precision weaponry to strike any selected target (we did not have this capability in Kosovo). (2) Because of unique national interests in Europe it is very difficult, and likely impossible, for our European allies to build such an umbrella themselves; moreover, their individual economies are much smaller and in many cases their technologies would not allow them to compete. (3) The United States should much more readily share its technology with its allies, and the rules governing such exchanges should be changed accordingly (many of the rules that determine technology transfer today are the same as they were twenty years ago). (4) We should *all* use commercial information, space, and communications technologies *much* more broadly. But there is no agency, no initiative, and as far as I can discern, no will to move aggressively to carry this out.

With these kinds of initiatives it should be recognized that the

United States will be the leader (but not the only player) in the development of RMA technologies (seeing the battlefield, communicating the information, and developing a range of smart longer-range strike weapons) and will be willing to *openly* share this information. The United States and NATO can work on the acquisition of ships, tanks, and aircraft to use the RMA technology and to "get the stuff where we need it when we need it," as well as to recruit and train men and women to fight future wars and participate in peacekeeping and peacemaking missions where necessary. This may sound simple, but it is not. Vision, insight into technology, openness to transatlantic business mergers, and political leadership and courage will be needed. Maybe our experience in Kosovo can be an incentive for implementing these kinds of changes. Without it, the U.S.–NATO "gap" is likely to grow, America could drift further from the European alliance, and the next Kosovo will be worse than this one.

THE DEBACLE OF TASK FORCE HAWK

The roots of the Task Force Hawk incident can be found in the age-old issue of military interservice rivalry and miscommunication coming back to haunt the United States in full view of our allies and adversaries alike.

For the United States and its NATO alliance partners in Kosovo in 1999, the gap between U.S. Army and U.S. Air Force doctrine led to a public embarrassment over the dispatch of advanced AH-64 Apache helicopter gunships to the air war over the Yugoslavia province.

Press accounts initially suggested that political opposition kept the Pentagon from using the low-flying tank-killing helicopters. But the issue was deeper and more troubling—the Army and Air Force had never thought about how the helicopters could be used in a high-technology air war.

First operational in 1986, the Apache is a 57-foot-8-inch-long attack helicopter capable of attacking and destroying a wide range of targets including tanks and armored vehicles. The two-man crew consists of a pilot and a gunner who can fire a range of lethal weapons including the laser-guided Hellfire air-to-ground missile, the unguided Hydra rockets, and a 30 mm cannon. During the Persian Gulf War, Apache units were equipped with the Pilot Night Vision Sensor, and a group of the attack helicopters fired the first shots of the war—infiltrating southern Iraq to destroy a key Iraqi air-defense radar site, thus enabling the first wave of attacking coalition aircraft to penetrate hostile airspace undetected.[21] Since the war, the Army has developed the AH-64D Apache Longbow variant, equipped with a millimeter-wave radar that provides the aircraft the ability to navigate in all kinds of weather and the ability to fire the Longbow Hellfire missile. The new Hellfire missile is a "fire and forget" missile that locks onto its target once fired, leaving the helicopter crew free to seek other targets or take evasive flight maneuvers.[22]

Two weeks after the NATO air campaign against Serbia kicked off on March 24, alliance officials revealed their intention to dispatch a force of AH-64 Apaches to Albania. The whole package included 24 Apaches, 18 ground-based Multiple-Launch Rocket System launchers, and a total of about 2,500 troops with 14 armored personnel carriers and other materiel.[23] Newspaper accounts reflecting the viewpoints of leaders of the American-led alliance portrayed the Apaches as vital to halting and reversing Serbian aggression. One account in mid-April stressed the AH-64's ability to engage enemy formations at close range:

> NATO plans to deploy them along with batteries of ground-to-ground mobile rocket systems capable of striking targets from a distance of 100 miles. Missile batteries would fire at enemy forces just ahead of the Apaches' arrival to force troops to duck and prevent them from firing at the helicopters. The Apaches' missiles can strike targets as far as five miles away. Though use of the Apaches involves greater risk than with higher-flying aircraft, they are not held back by darkness or the kind of bad weather that has hampered the air campaign so far.
>
> Since Yugoslav President Slobodan Milosevic's forces began driving ethnic Albanians from the Serbian province, advocates of more forceful action have looked on the Apaches, in conjunction

with A-10 "Warthog" ground-attack planes, as a way for NATO to sharply increase the pressure on Yugoslav troops.[24]

Eight weeks went by and the Apaches had not entered combat. It was not until the air war had passed its seventieth day that a clearer explanation emerged as to why NATO had approved the deployment of the Apaches but was unable to turn them loose against the Serbian Army and special paramilitary police in Kosovo. It turned out that the U.S. Army—which had designed and procured the gunships to fight closely together with its own ground forces—was not prepared to operate the Apaches as part of an Air Force-led aerial bombardment campaign in an environment where there were no friendly ground forces nearby. The dispute between the two services erupted in the pages of the trade journal *Defense Daily* on May 26:

> More than a decade after the Pentagon set out to bolster cooperation between the military services, a string of Air Force e-mail messages blasting the Army's plans to use AH-64 attack helicopters in Kosovo illustrates the problems still inherent in forging the type of united armed forces needed as the nation prepares to enter the 21st century, according to retired and active military officers and analysts.
>
> "The story here is that joint doctrine is a colossal failure," retired Air Force Col. Bob Gaskin, who served as the service's doctrine chief, told *Defense Daily* yesterday. "We've been working on joint doctrine for 15 years. The question to ask is what progress has been made? Smashing if you ask the bureaucrats, a colossal failure if you ask the operators."[25]

The dispute centered on the lingering disagreement between the Army and Air Force over the proper use of aircraft and helicopters on the battlefield—an argument that has raged since before World War II. The Army was reluctant to formally add the helicopters to the "air tasking order" that was controlled by the Air Force and that designated specific targets and aircraft used in the daily attacks. Officials told the trade journal that the Army was suspicious of handing over its valuable attack helicopters to a targeting process that the ground force leaders thought "a threat to battlefield initiative"—particularly since the air-tasking order is controlled by the U.S. Air Force and not the Army.

But the roots of the Apache snafu went deeper than mere human suspicion and bureaucratic rivalry. Designed and procured by the Army for the Army's use in ground maneuver campaigns, the Apaches—modern and lethal though they may be—are still unable "to integrate with support assets such as the Northrop Grumman E-8 JSTARS aircraft, the Lockheed Martin EC-130 Compass Call (radar) jamming aircraft and the Lockheed Martin F-16CJs equipped to defeat Serbian air defenses," the journal reported. Sixteen years after Grenada—during which Army ground troops found themselves unable to communicate with Navy carrier aircraft providing critical close-air support on the battlefield—the Army and Air Force assets rushed to Kosovo still could not communicate with one another.

The problem is rooted in the combat doctrine that defines the Army's design of weapons and the training of its forces in such a narrow way as to exclude Air Force systems that coexist in the inventory and on the battlefield, officials said. "What I'm saying is [the Army does] not have a good training tool to train and practice joint air integration," an Air Force officer in Albania wrote in an e-mail message obtained by *Defense Daily*. "It's a sad indictment that . . . the Army does not have a training method that forces them to integrate the other 'chess pieces' [e.g., Air Force weapons and systems] into their game besides the bag of pieces that they are already playing with," one Army official told the newspaper. General Bob Gaskin, the Air Force doctrine coordinator, acknowledged the lack of foresight in both services that led to the Apache fracas: "Joint air doctrine was envisioned only for the theater war air assets of each service. No one has ever seriously envisioned including Army aviation into a theater strategic air campaign, although doing so would be relatively easy if the mindset was there," Gaskin said. "Everybody trains, organizes, and equips to their service doctrine," he added. "When the services come to a war, they come with their service doctrines, not a joint doctrine."

The Task Force Hawk incident is a telling example of how the Revolution in Military Affairs remains hobbled by the paralyzed U.S. military bureaucracy and infrastructure, especially because the Apache is also a prime example of an advanced military combat weapons system that the "system of systems" architecture of the military revolution is designed to enhance and strengthen.

LESSONS STILL UNLEARNED

One month after the shooting stopped in the Balkans, Deputy Secretary of Defense John Hamre and Joint Staff Director Vice Admiral Vernon Clark appeared before the Pentagon press corps to discuss a planned review to "gather all of the lessons learned" from Kosovo. They announced a three-track approach to identifying shortcomings in the Kosovo campaign, including review groups focusing on (1) "deployment/employment" of combat forces, (2) the intelligence support for operations, and (3) issues pertaining to alliance and coalition warfare. The specific goal was to incorporate any needed changes into the new defense budget, which was scheduled to begin the cycle of congressional hearings in the spring of 2000. The nine-page transcript of the briefing contains several excellent and relevant issues that the Defense Department wanted to explore: How adequate is the fleet of specialized support aircraft such as the U-2 and EA-6B? (In fact, the Pentagon has already approved plans to create an additional EA-6B squadron to handle the intense demand from all of our military commanders.) How can the fleet of specialized aircraft manage a massive air campaign in a crowded airspace? How can it get vital intelligence data collected by "national means"—satellites and other sensors—into the cockpits of strike aircraft?

Those are good questions. But as I read the exchanges between these two leaders and the reporters who cover them daily, the real issue leapt off the page. It was what was *not* mentioned, what would *not* be examined or reviewed in the wake of Kosovo—the question of what to do with our military as a whole. The results of the Kosovo campaign reflected the enduring organizational flaws and internal paralysis that have hobbled previous American conflicts since World War II. The "victory" so quickly touted was at most a return to the status quo before the air strikes began (with the added irony that at year's end American and NATO peacekeepers were protecting Kosovar Serbs from Albanian ethnic cleansing). The U.S. military that has failed to implement the American Revolution in Military Affairs helped create a military and political quagmire that continues to this day.

And the wasteful impact of interservice parochialism, a budget fixated on "platforms" instead of information technology, and an unchanged military culture continue to erode our own military power.

CHAPTER 6

WINNING THE REVOLUTION

What if the United States could create a state-of-the-art military force from scratch? Suppose we could design a military force that would capitalize on our nation's technological strengths and operate free of the paralysis and parochialism sparked by entrenched interservice and interagency rivalries that have plagued the U.S. armed forces for most of the twentieth century? What would this new American armed force look like in the emerging era in which information technology is the defining factor in successful organizations, and in which industrial-era precepts of massed formations and rigid hierarchy are obsolete?

Such a force would be radically different from the current U.S. military in its most basic structure and operating characteristics. It would be a military force designed to exploit our lead in computer, sensing, and communications technologies in the hands of superbly trained American military personnel. It would be a force capable of achieving dominant battlespace knowledge and using it not only for general war but for maintaining peace. It would be a military that could, for the first time in history, pierce the fog of war, both minimizing American casualties, and *winning* the conflicts at hand. And waging war would be done for a cost of 20 percent less. This is the promise of the Revolution in Military Affairs.

The following are the basic outlines of the reinvigorated and modernized U.S. military that we can successfully create for the twenty-first century with the Revolution in Military Affairs.

- *Unified command structure.* Reflecting the constitutional requirements for civilian control of the military, we would retain a centralized command structure starting at the top with the civilian secretary of defense presiding over the Department of Defense and the armed services. Directly under the secretary would come the chairman, Joint Chiefs of Staff, and the Pentagon Joint Staff serving as the senior military leaders, transmitting the orders of the nation's civilian leaders to the regional military commanders in chief. The individual service military chiefs—the chiefs of staff of the Army and Air Force, the chief of naval operations, and the commandant of the Marine Corps—would each serve as the "chief executive officer" of his separate service, being responsible for administrative management of the huge facility infrastructure, human resources, and equipment. But the service chiefs, while retaining broad administrative authority as members of the Joint Chiefs of Staff, would relinquish the power to set individual service priorities and goals for weapons and equipment research and procurement. That critical function would be controlled by a single group reporting directly to the secretary of defense, composed of senior military and civilian Pentagon officials. The overriding purpose of the group would be to achieve a truly "joint" decision-making process for research and development priorities, and all procurement decisions.

  The Pentagon would still carry out military operations through a network of regional military commands oriented to strategically vital areas such as the Middle East, Europe, and the Asia-Pacific rim, and several multiservice headquarters charged with a "specified" mission such as transportation, space operations, or strategic nuclear weapons. And several new multiservice organizations that were created in the 1990s, such as a command responsible for defensive and offensive cyber-combat operations, would protect the military and civilian information technology networks from hostile attacks, and would conduct focused cyber-attack operations against the computer networks of terrorist groups or hostile governments.

- *Unitary military war-fighting organizations.* Although we would still have ground, naval, and air forces with individual service identities, the forces would be organized by what tasks they perform in battle rather than by what service branch they come from. Unlike the past tradition where the four services operated and maintained their own bases and installations, the basic building block for U.S. military force would be a range of *standing joint forces* that—regardless of their service composition or origin—would be located together, train together full-time, and deploy as a single entity. They could be general-purpose forces geared to land, air, and sea or they could be tailored to specific operating missions such as intelligence, surveillance, and reconnaissance; amphibious assault; long-range aerial strike; theater missile defense; sea control or armored ground assault.[1] These Standing Joint Forces organizations would be under the leadership of a three-star general or admiral and would number 15,000 to 50,000 personnel. There could be 25 or 30 of these forces in the U.S. military, each working for one of the regional four-star commanders in chief. *And* they would be organized for the most efficient and effective use of resources to accomplish America's missions around the world.
- *Embedded information warfare capability.* From the Joint Chiefs of Staff to the individual infantryman, sailor, airman, and Marine, the U.S. military force would define itself and its combat capabilities through superior information technology, enabling our fighting units to outmaneuver and outfight any conventional military opponent or irregular adversary. We would have an offensive capability to destroy the enemy force's intelligence and command-and-control system, and a defensive capability to protect the U.S. force from deception and attack by the enemy. All deployable combat units, whether tailored for ground, naval, or air operations, would be organized with a dedicated information warfare capability linking its commanders, weapons, sensors, and command-and-control systems.
- *From the command chain to the command network.* The industrial-age hierarchy from generals and admirals down to privates and seamen would be flattened significantly, and there would be fewer

intermediate layers of bureaucracy. In many places the chain of command would be replaced by secure and powerful networks that relay commands and critical battlespace information from the area of conflict to key decision makers, and from leaders and national intelligence agencies directly to the combatants, who will be able to respond more quickly and nimbly to directives from their commanders.

—*Lean and mean combat units.* Capitalizing on advanced computer processing power, networked sensors and communications, and long-range precision weapons, war-fighting units would be designed for maximum mobility, speed, and agility to be deployed to crisis areas from the continental United States and overseas bases.

—*Command, control, communications, computers, intelligence, surveillance, and reconnaissance capabilities (C^4ISR).* The wide range of early-warning, strategic intelligence, and tactical surveillance missions—the military's eyes and ears—would be supervised by a single military organization. Each combat organization would have a consolidated internal intelligence component at every key level in the chain of command, seamlessly connected to the command-level systems. Included would be tactical surveillance systems at the battlespace level whose data would be available from local units to senior leadership.

—*Consolidated global mobility.* Combat and support units would be designed and organized from the continental United States for rapid deployment worldwide with one organization being in charge of a modernized and expanded fleet of fast combat supply ships and military aircraft. Heavy equipment would continue to be prepositioned overseas at secure sites and onboard an enlarged fleet of cargo ships. Note that the United States could construct a number of huge Mobile Offshore Base deep-ocean structures that would anchor our military presence in trouble spots, providing basing for large logistics storage, several thousand combat troops, and up to several hundred tactical aircraft.

—*Consolidated advanced logistics.* The "tooth to tail" ratio of fighting versus support units would be twice what it is today. This leaner ratio of support to combat units would mark a revolutionary shift away from the bloated administrative and support structures of today, which constitute over 70 percent of the overall U.S. military in "tail," and only 25–30 percent in "tooth."[2]

—*Consolidated medical service.* Only one joint organization with doctors, administrators, and nurses of all four services involved would provide the best quality medical support to all men and women in uniform worldwide. This service could be under the executive management of one of the four services.

—*Homeland support.* Administrative and support functions such as personnel management, training, medical support, and maintenance would be organized by the CEOs of the four combat services (the chief of naval operations, the commandant of the Marine Corps, and the chiefs of staff of the Army and Air Force) in the continental United States with *joint* subfacilities at major overseas bases to provide complete support for members of all four services.

- *Cultural harmony.* Although the four armed services would retain their historical identities, their uniforms, and many of their traditions, the military as a whole would be trained and educated with a common joint military doctrine for combat and peacetime operations. Recruit enlisted personnel and officer candidates would be introduced to the military as a whole instead of just their own service. The various military schools would be united under a single organization, too. The existing service organizations that are responsible for formulating military service doctrines and strategies would likewise be unified (and streamlined), supporting and promoting a common vision of the U.S. military and leadership development in multiservice combat organizations instead of encouraging career paths dedicated to a solitary military service.

Do you recognize such a military force? Except for the fictional "mobile infantry" in Robert Heinlein's famed science fiction novel *Starship Troopers*,[3] no such military organization has ever existed. Neither the U.S. armed forces nor the military services of any other advanced nation today come close to the forces I've described here. But fielding such a force in the early twenty-first century is the logical objective and desired "cultural" goal if we are going to fully optimize America's military capability and take fullest advantage of the American Revolution in Military Affairs.

There has been some progress over the past five decades toward building a more efficient military—especially in the fourteen years since the passage, against the will of all levels of the Pentagon, of the Goldwater-Nichols Defense Reorganization Act in 1986. Improvements have come about as a result of enlightened and informed people recognizing the financial inefficiency and dangerous consequences of the "old ways." Since the early 1980s many military leaders, civilian defense officials, and their allies in Congress have helped achieve a degree of cooperation among the four combat services that few thought was ever possible. By the late 1990s it was not the work of isolated reformers but rather the avowed goal of the nation's senior military leaders to foster interservice cooperation and "jointness" as the tool to provide our men and women in uniform the technology, structure, and resources necessary to maintain the best military force in the world. But progress has been far too slow and only incremental; much more needs to be done. And time is running out as we squander enormous funds, while trying to retain the tradition and culture of the past. Often this foot-dragging is done under the premise that "our vital interests are at stake so we must be very cautious." My belief is that the riskiest approach is to continue on our present path.

SIGNS OF PROGRESS, SIGNS OF PARALYSIS

In the four years since my retirement from active military service, I have maintained a keen personal interest in the Pentagon's effort to implement the Revolution in Military Affairs. I have kept in touch with comrades still serving their country in uniform and have maintained a professional relationship with hundreds of military officers, Pentagon civilians, and members of Congress. Moreover, the Revolution in Military Affairs has been covered extensively in the military trade press, and the Pentagon itself regularly issues reports and news releases on various aspects of the revolution, including policy and budget documents (for a current reading and source list on the Revolution in Military Affairs, see the Suggested Reading list on page 241).

It is my sad judgment that the Revolution in Military Affairs is in serious trouble today. Efforts to enact reforms and to create new structures that will deliver on the promise of the information revolution have stalled as a result of the distraction of world events, poor planning, insufficient budget priority, and behind-the-scenes bureaucratic opposition to the dramatic organizational and cultural changes required to put these new technologies to work for us as fully as possible. And the current and looming Pentagon financial crisis makes the situation urgent.

It's not that the goal has not been identified. Indeed, before I left the service the Joint Staff and all of us on the Joint Chiefs worked on a thirty-four-page publication that articulated the philosophy and goals of the Revolution in Military Affairs that we in the Defense Department leadership had worked on during my tenure as vice chairman. The report, *Joint Vision 2010*, blessed and signed by General Shalikashvili, the chairman of the Joint Chiefs of Staff, presented the case for a highly complex transformation of the U.S. military in remarkably clear language. Its priorities include:

- *Common direction.* The report, *Joint Vision 2010*, constitutes "a common direction" for the four uniformed services to develop their unique capabilities "within a joint framework of doctrine and programs." The four operational concepts that will guide all of the services in future war fighting—"dominant maneuver," "precision engagement," "full dimensional protection," and "focused logistics"—are more or less the themes we drafted for the Revolution in Military Affairs.
- *Seamless integration of the four services.* The report calls for the Army, Navy, Air Force, and Marines to cooperate at every stage of military life, from training to combat. "Simply to retain our effectiveness with less redundancy, we will need to wring every ounce of capability from every available source," the report states.
- *Capitalizing on new technology.* At the heart of future U.S. military combat superiority, the report acknowledges, will be long-range precision weapons and a variety of enhanced delivery systems. Other technological gains such as new forms of stealth technology, nonlethal weapons, and information systems also promise to make the U.S. military force more flexible and more powerful, the report states.

- *Recognizing the importance of information superiority.* The very process of collecting and processing information through the power of computers and global satellite communications will become a prime objective in future warfare. The U.S. military will develop doctrine, tactics, and weapons to conduct both offensive and defensive information operations.
- *Timeline.* Full integration of the "system of systems" that will leverage the U.S. military into the twenty-first century is expected by the year 2010.

Joint Vision 2010 and a follow-up report, *Concept for Future Joint Operations—Expanding Joint Vision 2010,* described the goals set out by Secretary Perry, Undersecretary Deutch, General Shalikashvili, and me four years ago when we first sought to bring about the Revolution in Military Affairs. These goals have been echoed since then by Secretary of Defense William Cohen, who replaced Perry in January 1997, and Joint Chiefs Chairman General Henry Shelton, who succeeded Shalikashvili. In his introduction to the fiscal year 1999 *Annual Report to the Pentagon and to the Congress* released in May 1998, Secretary Cohen summarized the Pentagon's goal: "We will execute the [U.S. military's twenty-first century] strategy with superior military forces that fully exploit advances in technology by employing new operational concepts and organizational structures. And we will support our forces with a Department that is as lean, agile, and focused as our warfighters."

On the Revolution in Military Affairs Cohen was even more explicit:

> The Department's willingness to embrace the Revolution in Military Affairs (RMA)—to harness technology to ultimately bring about fundamental conceptual and organizational change—is critical at this stage of the [Pentagon's] transformation strategy.
>
> Today, the world is in the midst of an RMA sparked by leap-ahead advances in information technologies. There is no definitive, linear process by which the Department can take advantage of the information revolution and its attendant RMA. Rather, it requires extensive experimentation both to understand the potential contributions of emerging technologies and to develop innovative operational concepts to harness these new technologies.[4]

The good news is obvious: the U.S. military leadership philosophically supports the changes necessary to move the American armed forces into the twenty-first century. Moreover, each service (individually!) has launched a number of experiments and initiatives to adapt computer applications and other new technological applications to its particular force structure. On the surface, there seems to be steady progress.

The Navy is pressing ahead with a number of RMA initiatives through its ongoing Fleet Battle Experiments, including expansion of the *Smart Ship* program aimed at sending out smaller crews while using automation and computers to keep their edge in warfare. In the *Ring of Fire* concept, under Navy study since 1997, the Navy envisions a computer *wide area network* that would link all of its ships and naval aircraft in a battle group together, but the network could be expanded to bring in Army and Marine Corps artillery units that are already ashore. The carrier force has experimented with *surge exercises* with the aim of enabling the carrier air wing to fly more combat missions by managing the number of aircraft on each ship and increasing the number of pilots assigned to the ship.

Marine Corps leaders, mindful of their tragic experiences in Beirut in 1983 and the Army Special Forces actions in Somalia a decade later, have conducted major training exercises to prepare their units for the harsh realities of urban fighting. "If there is an enemy out there who wants to make a difference, he can only make a difference by getting us into a complex, chaotic, deadly environment that negates our technology and our strength and capitalizes on their strengths," then Marine Corps Commandant General Charles Krulak noted recently. "It's called the cities."[5]

So, too, the U.S. Air Force has sought to increase its combat capability through developing new weapons and tactics. The air service in 1999 announced it was expanding the concept of the *Air Expeditionary Wing*—consisting of a team of bombers, tankers for refueling, strike fighters, and air defense aircraft that train and deploy together. The Air Force is even studying the urban combat terrain to find ways in which its aerial combat units can protect the U.S. force and defeat an enemy entrenched in the "vertical" battlefield of buildings and street systems, using a combination of space satellites, and advanced ISR systems to locate the foe and help defeat it. In 1999 Major General Norton Schwartz, the Air Force director of strategic planning and the chairman of the Defense Department's Joint Urban Working Group, said that the Air Force had begun researching ways it can provide the U.S. forces commander essential information about the "three-dimensional battlespace" and identify key "nodes" that support the adversary and then attack them using precision weapons, nonlethal systems, and information warfare. "Failure to use aerospace power in urban environments puts American success at risk," Schwartz said. "Don't go downtown without us."[6]

So far, so good. But when you read the fine print of the various Defense Department reports, there is cause for serious concern. Let's look at the Army's efforts to build a new force around the potential of "digital information."

FORCE XXI: THE ARMY GOES DIGITAL

The Army has spent most of the 1990s preparing for war in the digital era. In 1992 the service undertook a wide-ranging series of studies and "battle laboratory" experiments with the goal of designing the army of the next century. The goal: the *Force XXI* program (for "Force Twenty-first Century," of course), which aims to apply advances in computer processing power to ground combat operations by 2010; and the *Army After Next,* which is meant to replace Force XXI in 2025. Secretary Cohen in his 1998 annual Defense Department report underscored the Army's commitment to fielding an RMA-ready ground force:

> The Army is identifying new concepts of land warfare that have radical implications for its organization, structure, operations, and support. Lighter, more durable equipment will enhance deployability and sustainability. Advanced information technologies will help the Army conduct rapid, decisive operations. The force will be protected by advanced but easy-to-use sensors, processors, and warfighting systems to ensure freedom of strategic and operational maneuver. . . . The Army will require flexible, highly tailorable organizations—from small units to echelons

above corps—to meet the diverse needs of future operations and to reduce the lift requirements for deployment.[7]

In order to modernize, the Army plans to retool the division, its basic combat unit for most of the twentieth century. At present, a division includes between 15,000 and 18,000 soldiers and includes

- A division headquarters for the two-star commanding general.
- Three ground combat brigades of about 3,000 personnel each, consisting of three infantry, mechanized, or tank battalions of about 750–800 troops designed to attack the enemy and seize and hold territory. Each battalion is organized into three companies of three or four platoons apiece.
- One artillery brigade—tanks, armored carriers, and mobile artillery—to provide fire support for the ground troops.
- One aviation brigade to provide attack helicopters for close air support, transport helicopters to move troops and supplies, and scout helicopters to conduct reconnaissance.
- One engineer brigade to provide combat engineer support, through constructing bridges or fixed defenses, for example, or by clearing paths through minefields.
- One brigade-sized division support command to provide logistics, medical, and other services to the combat units.

As the 1990s began, the Army had eighteen such divisions. But budget cuts during the Bush and Clinton Administrations meant that the Army had to deactivate eight divisions—nearly 40 percent of its Cold War-era fighting force. Needless to say, Army leaders were concerned about that. What if a full-scale war broke out? So they sought to identify how the Army could apply information-age technology in order to increase each division's combat power, even as the division was reshaped into a slightly smaller, more mobile force. The goal, announced in 1994 by then Army Secretary Togo West and Chief of Staff General Gordon Sullivan, was Force XXI.[8]

It was slow going at first, but by 1999 the Army had refined plans to "digitize" its four corps headquarters and eight divisions in the first decade of the twenty-first century, with the first "digitized" division due to be in place by the end of 2000.[9]

The "digitized division" concept was tested in the Division Advanced Warfighting Exercise held at Fort Hood, Texas, in November 1997.

There the 4th Infantry Division and its brigade commanders were forced to respond to a computer-generated campaign in which the experimental division fought an enemy field army equipped with military systems and weapons anticipated to be on the world arms market in the year 2003. Following is the organizational outline of the Army's "new" approach to warfare, including the communications, computer systems, and weapons that have been added to the "experimental" division, and a reading on the performance of this approach in the field:

- *Force organization.* The "experimental" division, like its traditional predecessors, was built around three ground combat brigades of three battalions per brigade, but each had fewer tanks, armored personnel carriers, and other vehicles. The new division had reduced artillery systems—only 54 self-propelled artillery pieces, a 25 percent cut—but twice as many Multiple-Launch Rocket System launchers.[10] Combat support was cut 21 percent, from 5,500 to 4,300 soldiers, on the assumption that the better support people can communicate, the fewer of them are needed to do their work.
- *Computer systems.* The new division design embraced eighteen separate existing and experimental digital computer and communications systems, including the existing Army Tactical Command and Control System suite deployed in the Army in the 1980s. This existing system includes the Maneuver Control System; the All-Source Analysis System for intelligence data; the Advanced Field Artillery Tactical Data System for directing artillery fire; the Combat Service Support System for coordinating resupply; and the Forward Area Air Defense Command, Control, and Intelligence system for protecting units from enemy air attack.

  New experimental systems tested during the exercise included a secure video teleconferencing network allowing the division commander and his subordinates to plan operations without the delay and danger of mustering at a single site for in-person meetings.

Another new system employed was the Raptor Intelligent Combat Outpost, a combination of sensors, communications gear, and antiarmor mines that can be deployed swiftly to detect and halt the enemy's advance. A new Analysis and Control Team Enclave, serving each ground maneuver and artillery brigade commander, can provide instantaneous reception and analysis of imagery of enemy ground forces from the JSTARS surveillance aircraft, as well as from various unmanned aerial vehicles that can reconnoiter the battlefield. A new digital Air and Missile Defense Planning and Control System can provide the division commander a three-dimensional situational awareness picture of the extended battlespace to show enemy missile threats. Once the division commander sees this, he can decide whether to launch artillery, move his troops, or follow some other strategy. And a prototype Digital Topographic Support System can provide generated digital data on the battlespace terrain and instantaneous mapping.

- *Weapons and field equipment.* The digital division was assumed to be equipped with numerous state-of-the-art weapons systems that are in actuality still under development or just now being produced for service. Included were the RAH-66 Comanche scout helicopter, the planned XM2001 Crusader self-propelled field artillery system, and the AH-64 Apache Longbow (equipped with advanced millimeter-wave radar for all-weather targeting of "fire and forget" Hellfire air-to-ground missiles).
- *Exercise performance.* Given the task to defeat a larger conventional force on a battlefield spanning an area of 285 by 156 miles, the Force XXI exercise judges concluded that the new digital division succeeded in defeating the enemy. "Digitization enabled the [experimental division] commander to accept the risks of fighting without an operational reserve and with greater gaps created by the extended battlespace.[11]

It appeared that the Army was leading the way toward the Revolution in Military Affairs, investing about $2.8 billion a year in the various technological applications and war-fighting experiments.[12] But a closer look shows that the Army hadn't changed its approach nearly as radically as I am contending it must.

FIELDING A MORE FLEXIBLE FORCE

The main criticism from experts both within and outside the Army is that initiatives such as Force XXI and the Army After Next are still wedded to the corps and division structure that were invented by Napoleon and took their current form between the two world wars. In a groundbreaking study of the U.S. Army, *Breaking the Phalanx: A New Design for Landpower in the Twenty-first Century,* Colonel Douglas A. Macgregor wrote in 1997, "Proportionally, today's Army force structure is still composed of large industrial age combat forces capable of massing firepower."[13] The basic triangular shape of the division (three platoons per company, three companies per battalion, and three maneuver brigades to the division) have been left intact. Macgregor and other experts point out that the Army is trying to graft new technology onto an obsolete organization and culture in the belief that the technological gains will correct the force's structural flaws.

As an active duty officer when he wrote his book (as he remains today), Colonel Macgregor committed a bold act of intellectual leadership—bordering on heresy—by recommending that the Army abolish the division concept altogether. In its place, he argued for a digitized U.S. Army formed around twenty-six "mobile combat groups."[14] But he recognized that in peacetime this was a hard argument to put across. "Because the character of the threat is no longer specified, it is not surprising that the Army's Force XXI program has not resulted in any significant change in the warfighting structure of Army forces since Desert Storm," Macgregor wrote.

Macgregor argues that technology alone cannot provide the major advance in U.S. military capability that the information revolution makes possible. Rather, he stressed "the importance of the right organization for combat within a coherent doctrinal framework. . . . The evolution of the United States Army into a new form will depend on more than the incorporation of new technology."[15]

Jeffrey R. Cooper, director of the Center for International Strategy and Policy at Science Applications International Corporation, has also warned of the fallacy of merely laminating new technology onto obsolete military organizations. "These benefits [from the computer revolution], however, cannot be captured by piecemeal adoption of technologies; they demand fundamental alterations in the concepts for prosecuting the operational level of war," Cooper wrote in a 1996 essay.[16]

To its credit, the Army has acknowledged that its planned division design under Force XXI is simply too slow and too bulky to be of much use in future wars. In a 1997 computer war game Army experts tested a variety of military threats the United States might face twenty years from now to see how Force XXI and Army After Next forces might respond to them.[17] One was a terrorist bombing in an overpopulated city center like Lagos, Nigeria; another was a chemical weapons attack in a First World capital; another was a massive mobilization of a poorly equipped but ardent peasant army such as might occur in Pakistan or North Korea. The analysts conducting the exercise found that "strategic speed"—the pace at which Army forces can deploy from the continental United States to a foreign trouble spot—will constitute "the dominant factor at the strategic-political, strategic-military and operational levels of war."[18] In other words, the early bird wins the war. But in a refreshing display of candor, the report concluded:

> Concerns emerged during the (war) game over an obvious disparity between the strategic speed of an AAN (Army after Next) force—arriving from CONUS [continental United States] ready to fight within 48 hours—and the follow-on CONUS-based Army XXI force. To allow the ability both to pre-empt an enemy from setting his force in a theater and to continue unrelenting sustained pressure over time, the projection schemes of both (U.S.) forces should be seamless and firmly joined. It became clear during the game that by 2020 a mature Army XXI force must be much more projectable than heavy forces are today, inferring perhaps the requirement to move globally from a staging point to a distant battlefield in no less than two weeks.[19]

But as the Army itself knows, the "digital" division design now being put in place at Fort Hood is still a very heavy and bulky collection of hardware and equipment. Even this "streamlined" division has 608 tanks, combat infantry vehicles, motorized rocket launchers, and the like; it is not going to travel anywhere fast—not today, and not in the next decade.

So in late 1999 the Army launched a third reform initiative aimed at achieving a compromise between ground combat power and mobility. General Eric Shinseki, the new Army chief of staff, announced in late October 1999 that the service would create five new "medium" brigades that would be armed and equipped with a family of lightweight wheeled combat vehicles such as the Canadian Light Armored Vehicle that is in use by the Marines.

In a major speech to the Association of the U.S. Army, Shinseki said a heavy armor unit at Fort Lewis, Wash., would be converted by the summer of 2000, and a total of five brigades would trade in their heavy 1980s-era M1A1 Abrams tanks and M2 Bradley infantry fighting vehicles by 2003. And, he indicated, the entire U.S. Army will be drastically redesigned by 2020.[20]

Army officials say the imperative is to preserve troop protection (with armored vehicles) as well as combat punch while fielding lighter, more easily transported equipment. The sheer weight of current Army systems is the pivotal issue in this ongoing effort, officials said. The Air Force's new C-17 transport plane can carry only one 67-ton M1A1 tank at a time, but it can ferry five of the 24-ton LAV-25s and their crews over the same distance.

Behind the Shinseki initiative—which merits strong support from the Pentagon leadership and Congress—is another dreary narrative of stifled innovation. Twenty years earlier the Army had wrestled with the same dilemma when it recognized the growing security threat in the Middle East and ordered the 9th Infantry Division at Fort Lewis to restructure as the only "motorized" infantry division in the service. This unique design envisioned a light armored howitzer and fast off-the-road vehicles to ferry infantry troops, but after six years of experimentation, the division was practicing war games with unarmored "Humvee" vehicles that provided soldiers no protection from enemy artillery or even rifle fire. The lightweight mobile gun at the heart of the motorized infantry concept was never purchased, and the 9th Division became one of the first units to be deactivated as the ground service began its post-Cold War retrenchment from eighteen to ten divisions.

I deem it a hopeful sign that the current push for a "medium" brigade strike force has the clear support of the Army's top leadership and is being managed by all of the service combat and support branches, in contrast to the 9th Division, which was quietly derided by traditionalists in the Army as an unworkable concept.

Meanwhile, the service is still wrestling with its two main modernization efforts—Force XXI and the Army After Next. I fear that two of the Army's premises here are likely to hamper—if not doom—its efforts at transforming the force.

First, the Force XXI plan upgrades the Army forces in precisely the wrong order. The decision to "digitize" the U.S.-based heavy divisions first, then the first-to-fight "light" divisions, and finally the overseas-based American units means that the Army units most likely to face combat in the decade ahead—the "light" forces of the Army's XVIII Airborne Corps and units overseas in Europe and Korea—won't be upgraded until 2009, and until then they will remain as they are, essentially unchanged since Operation Desert Storm. That process is like trying to enter virtual reality while using a DOS computer with two floppy disks and no hard drive.

Second, the Army seems to have suffered a failure of nerve, because it is attempting to graft the information-age technology on George Marshall's pre–World War II division design, and worse, is making little effort to integrate the new design with the other services. I believe the service should pay heed to its critics who warn that technology misapplied within an organization only guarantees failure.

RHETORIC AND REALITY

When the Joint Chiefs of Staff approved the final draft of *Joint Vision 2010* for publication in July 1996, stating their long-term goal of revitalizing the four services through the Revolution in Military Affairs, they inserted a carefully written caveat at the end. I vividly remember how many contentious meetings we spent drafting this particular warning, which was boiled down into a short passage in the thirty-four-page document. Here, in military parlance, is what we said:

> As we implement this vision, affordability of the technologies envisioned to achieve full spectrum dominance will be an important consideration. *While we anticipate that some significant improvements in capability may be gained economically, for example through dual-use technologies for C⁴I (command, control, communications, computers, and intelligence) others will be more difficult to achieve within the budget realities that exist today and will exist into the next century* [italics added]. . . . Ultimately, we will have to measure continuously *the affordability* of achieving full spectrum dominance against our *overarching need* to maintain the quality of our forces, their readiness, and the force structure needed to execute our operational tasks between now (1996) and the year 2010.[21]

The translation is simple and stark: We need to upgrade our military, and fast, but we can't do it without money. Even as we built a consensus on what was needed to jump-start American military power into the twenty-first century, we knew that the White House and many congressional leaders were bent on drastically reducing defense spending, whatever the cost to our national security. Our long-term research-and-development projects were falling farther and farther behind. The existing force was being run ragged by the ceaseless humanitarian missions, peacekeeping assignments, and low-intensity conflicts the Clinton Administration had involved us in throughout the world. Because the White House refused to meet the minimum budget requirements for the Defense Department, the military services had to repeatedly raid their "investment" budgets—financial accounts for long-term modernization of ships, aircraft, and combat equipment—simply to keep the force moving in current operations. The American Revolution in Military Affairs appeared bankrupt before it had gotten started.

The Illusion of Jointness

That said, it is important to add that the military shares the blame for its own abysmal lack of progress. By failing to work jointly to apply the new technology, the four services are digging themselves four separate, low-tech graves. A prime example is the call for "joint doctrine" among the four armed services—an ideal espoused by senior U.S. military leaders since 1986 under the impetus of the Goldwater-Nichols Defense Reorganization Act. The legislation has generated protracted discussion and debate, led to reams of formal studies, prompted the creation of a new military doctrine command—and as yet joint doctrine has failed to materialize.

Doctrine is a basic military term defined by one expert as "the combination of principles, policies and concepts into an integrated system for the purpose of governing all components of a military force in combat, and assuring consistent, coordinated employment of these (military) components. . . . Doctrine represents the available thought on the employment of forces that has been adopted by an armed force."[22] As we mentioned earlier, misunderstandings between or among the different armed services as to how they plan to fight a battle can lead to serious blunders, friendly-fire incidents, or even defeat.

Standing Still Is Not an Option

The United States does not have a monopoly on the Revolution in Military Affairs, and there is plenty of evidence that other nations—including those that are not close allies and some that are unfriendly to us and our international interests—are keenly studying contemporary military trends with a view toward advancing their capabilities and their strengths.

The sudden nuclear arms race between India and Pakistan in 1998 is one manifestation of two regional powers striving to break out of the old military balance using a technological breakthrough to boost their power (although those two nations have essentially opted to move from classic Third World military force to regional strength with existing weapons of mass destruction). The ongoing effort by the Swedish military to transform itself into a modernized, technologically sophisticated force is another sign. But it is the People's Republic of China that presents us with the starkest example of why we cannot afford to rest on our current military lead.

In 1999 a stark revelation by the bipartisan congressional Cox Committee investigation showed China's intention to forge its own Revolution in Military Affairs. The committee's report disclosed a massive, ongoing Chinese espionage and technology-transfer effort to secure advanced U.S. nuclear warhead designs, ballistic missile components, and supercomputers.[23]

Obscured by the panel's alarming discoveries of lax physical security in the U.S. Department of Energy nuclear laboratories, was a detailed account of China's ongoing effort to forge its own military revolution. It is worth retelling here in some detail. The report stated:

> In 1996, [Chinese] "Paramount Leader" Deng Xiaoping adopted a major initiative, the so-called "863 Program," to accelerate the acquisition and development of science and technology for the PRC [People's Republic of China]. Deng directed 200 scientists to develop science and technology goals. The PRC claims that the 863 Program produced nearly 1,500 research achievements by 1996 and was supported by nearly 30,000 scientific and technical personnel who worked to advance the PRC's "economy and . . . national defense construction."[24] The most senior engineers behind the 863 Program were involved in strategic military programs such as space tracking, nuclear energy and satellites. Placed under COSTIND's [the State Commission of Science, Technology and Industry for National Defense] management, the 863 Program aimed to narrow the gap between the PRC and the West by the year 2000 in key science and technology sectors, including the military technology areas of: astronautics, information technology, laser technology, automation technology, energy technology and new materials. The 863 Program was given a budget split between military and civilian projects, and focuses on both military and civilian science and technology. The following are key areas of military concern:
>
> - Biological warfare: The 863 Project includes a recently unveiled plan for gene research that could have biological warfare implications.
> - Space technology: Recent PRC planning has focused on the development of satellites with remote sensing capabilities, which could be used for military reconnaissance, as well as space launch vehicles.

- Military information technology: The 863 Program includes the development of intelligence computers, optoelectronics, and image processing for weather forecasting; and the production of submicron integrated circuits on 8-inch silicon wafers. These programs could lead to the development of military communications and intelligence systems; command, control, communications, and intelligence systems; and advances in military software development.
- Laser weapons: The 863 Program includes the development of pulse-power techniques, plasma technology, and laser spectroscopy, all of which are useful in the development of laser weapons.
- Automation technology: This area of the 863 Program, which includes the development of computer-integrated manufacturing systems and robotics for increased production capability, is focused in the areas of electronics, machinery, space, chemistry and telecommunications, and could standardize and improve the PRC's military production.
- Nuclear weapons: Qinghua University Nuclear Research Institute has claimed success in the development of high-temperature, gas-cooled reactors, projects that could aid in the development of nuclear weapons.
- Exotic materials: The 863 Program areas include optoelectronic information materials, structural materials, special function materials, composites, rare-earth metals, new energy compound materials, and high-capacity engineering plastics. These projects could advance the PRC's development of materials, such as composites, for military aircraft and other weapons.[25]

The Cox Committee also noted that in 1996 the PRC government announced the formation of the "Super 863 Program," a major follow-on to the ten-year-old military technology acquisition effort to coordinate and promote advanced technological development between the years 1996 and 2010. Its field of activity includes original research, exploitation of dual-use technological applications introduced into China by joint ventures with foreign companies, information obtained from open sources, and information covertly acquired through espionage and other means.[26]

There is no mystery in China's long-term aspirations. As early as

1998 the PRC Defense white paper stated that "no effort will be spared to improve the modernization of weaponry," and the Cox Committee assessment of the specific priorities for Chinese military modernization include battlefield communications, reconnaissance, space-based weapons, mobile nuclear weapons, attack submarines, fighter aircraft, precision-guided weapons, and rapid-reaction ground forces.[27]

We have already seen signs that China is racing to deploy a new generation of intercontinental ballistic missiles, including mobile systems that can make retaliatory counterstrikes difficult to succeed. The People's Liberation Army did not stop with its aggressive missile tests in 1996 that overflew Taiwan, but has reportedly deployed dozens of medium-range missiles along its eastern coastline that directly threaten Taiwan. The Chinese Air Force has acquired Soviet-era MiG-29 and Su-27 combat aircraft from Russia and is apparently girding to begin mass production of those modern designs, and the Chinese are beginning to deploy aerial refueling tankers that can extend the range of military aircraft to provide a regional power-projection capability.

And Chinese aerospace officials have made no secret of their ambitions to vault the People's Republic into the new century as a major power in space. As early as 1998 officials said they were planning to launch a constellation of four imaging satellites and two radar satellites within the next few years. As one influential Asian newspaper noted:

> What is remarkable is that China is following the U.S. and Russian example with gathering speed. China's new satellites represent the active element of an intensive Chinese effort to use space for military missions as well as to deny use of space to its enemies. China's satellite and anti-satellite efforts, in turn, are the manifestation of a broader developing information warfare doctrine that seeks to protect and exploit its sources of electronic information while denying or attacking enemy information systems. China's military planners understand that future battles will be determined by control of information and that space is the new high ground for information control and terrestrial military dominance.[28]

I agree with analysts who say there is no need for immediate panic over the extent of technological applications and hardware that China has acquired from the United States and other Western nations. In 1999 retired Admiral David Jeremiah, friend and predecessor as vice chairman, led an independent review of the Cox Committee findings and concluded that although the espionage has helped China accelerate the pace of its nuclear weapons program and other projects, the espionage "has not resulted in any apparent modernization of their deployed strategic force or any new nuclear weapons deployment."[29]

It is not the military capability of China today that should cause us concern. Rather, it is the clear message that Chinese government officials are sending—amplified by their ambitious and wide-ranging effort to acquire modern military technology—that portends a significant Chinese military threat to the interests of the United States and its East Asian allies a decade from now. And not just in East Asia: Chinese proliferation of ballistic missile technology to Pakistan, Iran, and other nations is hastening the day when the United States will face ballistic missile threats to our forces in the Eastern Hemisphere that we have never had to worry about before. Progress has been made since the spring 1999 release of the Cox Committee report to close off export loopholes and tighten up on our nuclear laboratory security systems that enabled China to acquire this invaluable information. But it is obvious that China has dedicated itself to its own military revolution aimed at meeting our current capabilities. We have no alternative but to press on with the American Revolution in Military Affairs to forestall the day we are checkmated by the rise of another military power whose interests clash with our own.

THE REVOLUTION'S NEXT PHASE

During the 1990s we succeeded at building the first phase of the American Revolution in Military Affairs. We built and fielded the core information-collection systems, the communications systems, and most of the precision-guided weapons that the U.S. military will require, with the remainder still in research and development. We are well on the path to the second wave of the emerging "system of systems"—a higher level of systems integration that constitutes the heart of the military revolution.

The next phase will focus on *systems integration*, the process by which we tie these new military technological applications into the organization and force structure of the U.S. military. Systems integration is not a difficult concept: it involves getting various components of the "system of systems" to work together better. By bringing the commercial dynamism of information management and communications into the military, we are clearly moving to a higher order of understanding of the battlefield. This will mean some subtle but important and difficult adjustments in our priorities for technological development and military procurement, for it implies that we should not buy new equipment unless it meets tougher standards of cross-service interoperability. We need to agree in advance that all new equipment—weaponry, aircraft, computers, communications devices, you name it—can be used by all four service branches and that we will look very carefully to be sure that we are not buying redundant systems. And with regard to technological development and the trade-off with traditional "platforms," we need to make adjustments to many long-standing procurement goals.

For example, in developing future tanks, combat aircraft, and naval warships, we would place new emphasis on making these platforms capable of transferring battlespace knowledge among the forces in addition to continuing to make these platforms faster, better protected, and longer operating. That sort of adjustment would, in turn, mean that the procurement of the next tank, fighter, or cruiser would have to be far more "information oriented" and synchronized across the four military services than is the case now. We would begin to assess development and procurement options differently, evaluating the competing candidates in terms of the extent to which they enhance each other. We would begin to see aircraft, or tanks, or submarines increasingly as only components of a joint warfighting system and not as weapons "platforms" assessed in isolation of the rest of the force. We would increasingly evaluate them not only against the standards each of the military services has used in the past, but in terms of how they contribute to obtaining dominant battlespace knowledge and together winning the battle or achieving the nation's goals. Likewise, this orientation would alter the way we spend research and development dollars and what we demand of our research and development community. We would place higher priority on better integrating data collection with communications, on better integrating communications with precision weapons, and on better integrating precision force use with battle damage assessment capabilities.

Unless the Pentagon budget rises dramatically—and this is simply not going to happen—this shift will have to be paid for with funds that otherwise would be spent on the traditional research and development "requirements." It will also almost certainly mean much greater interest in drawing from the commercial sector rather than the government laboratory and the traditional defense contractor base that we built over the last half century to wage the Cold War, because the cutting edge of information technology increasingly emerges from the commercial marketplace.

Completing the Revolution in Military Affairs will involve the controversial and difficult effort of integrating innovative military technological applications—and organizing new ways of conducting warfare—into a realigned military force structure. Identifying specific organizational reforms will be hard and difficult work, because *every change* will challenge and threaten a host of entrenched military traditions and bureaucratic interests.

It is important to grasp one root concept that is necessary for true reform: to choose *synergy* over *specialization.* Today joint military operations are taken as a given in Pentagon policy planning, and it is virtually impossible to find anyone who professes to be against them. But the unanimous support belies a continuing serious disagreement as to the exact nature of a joint operation. According to the "specialization" view, a joint operation is one that deploys the "best qualified" force component—such as a Marine Expeditionary Unit versus an Army Ranger battalion—for a given mission, regardless of which service branch the force comes from. According to the "synergy" view, the military should try to combine forces whenever possible in order to ensure a mission's success.

When discussing joint operations, military people commonly speak of the "toolbox"—a collection of different military forces a commander assembles in the way he believes can best perform the mission he is assigned. An advocate of the specialization view would say that the joint commander has to choose the right tool for a particular job. For instance, if he is required to plan and conduct a strategic bombardment campaign, the commander ought to assign this to the U.S. Air Force, the component that knows the most about strategic bombardment. An advocate of synergism would argue that the joint task force commander ought to use several tools together, say, by assigning Navy carrier aircraft to suppress enemy air defenses and clear routes for land-based Air Force bombers, or intermingling the strike assets of different services.

Although both of these force-building models aim at jointness, "specialization" and "synergy" have starkly different implications for the future of the military. The essence of specialization—the traditional approach employed by the U.S. military since World War II—is to clearly differentiate combat responsibilities among different military service organizations (Navy carrier aviation versus Air Force units, Marine Corps amphibious forces versus Army "light" divisions). The synergy approach assumes that smaller elements of the different military services—such as Navy EA-6B Prowler electronic warfare aircraft and Air Force F-15E Strike Eagle all-weather attack aircraft or Army Apache helicopters deployed on ships with Marine Expeditionary units—can and should be used together when the mission requires.

Specialization reflects the long and inbred preference within the U.S. military to *support one's own service above the others.* It takes advantage of inherent efficiencies in the integrated traditions, doctrines, discipline, and procedures of a single service and not of joint force. A good example of how specialization dominated a military campaign occurred during the Vietnam War, when the U.S. Air Force operating from bases in Thailand and the U.S. Navy operating from aircraft carriers in the Tonkin Gulf arbitrarily divided North and South Vietnamese airspace into "route packages" (or in the case of Desert Storm, the assignment of aircraft strike packages to the Navy and Air Force)—not out of any effort to maximize the use of military power against the enemy, but out of a need to avoid inadvertent conflict by two separate air arms that could not effectively communicate with one another and whose tactics did not mesh. The current American Revolution in Military Affairs inevitably is grounded in synergy, because the "system of systems"—in command-and-control, communications, intelligence gathering, and firepower—inevitably works to break down artificial hierarchies and bureaucratic walls to find the most efficient combination of weapons, systems, and people. My experiment as 6th Fleet commander in 1992 to strip the flight deck of the USS *Saratoga* (strongly opposed by senior Navy admirals) to accommodate Army special operations helicopters was an effort of synergy that a decade earlier was impossible to imagine, much less carry out.

But any study of U.S. military operations from Vietnam up through the present day reveals that there have been far fewer "joint" operations than Pentagon rhetoric suggests. One reason for this is that the educational and training regimes within each service continue to focus on preparing officers and enlisted personnel to master their own narrow specialties at the expense of gaining a perspective on the common defense.

The logic of the Revolution in Military Affairs points to one root concept that the U.S. military cannot avoid: the need to organize standing joint forces that train and operate together full-time. Each of the military services alludes to joint operations. When the U.S. Air Force describes its powerful concept of *Global Reach, Global Power* it acknowledges that while its land-based bomber forces can reach out from the United States to strike anywhere on earth, the tactical airpower that can supplement them will come from forward-deployed aircraft carriers. When the Navy describes its concept of projecting power from the sea to the land, it acknowledges how Air Force aerial refueling tankers can make the Navy's power projection capabilities stronger and how Air Force bombers truly bring the "mass" to a bombing campaign. Since the 1970s the Army has talked and written of coordinated ground and air operations, even coining the concept of the *AirLand Battle.* Both the Army and Marine Corps acknowledge the need for airpower to supplement the ability of their rapid deployment units to sustain their operations effectively.

But where do we put this rhetoric into effect? The answer is clear: create standing joint forces that will actually deploy and fight the enemy. Even now, nearly fourteen years after the Goldwater-Nichols Defense Reorganization Act formally mandated joint military operations, most of the military combat units in the U.S. armed forces still reside at their own bases in the continental United States. They come together with forces from other services only in very small numbers and in large annual field training exercises. Likewise, most of the combat organizations overseas come together only at the last minute with their counterparts in an emergency or actual combat (and then they often find their level of training or the equipping of common communications, fusion centers, and sensors to be different and lacking).

Although much remains to be done, there are some encouraging signs of progress in this vital area. In 1999, under an initiative directed by Defense Secretary Cohen and the military leadership, the multiservice U.S. Atlantic Command formally renamed itself the U.S. Joint Forces Command as a symbol of its new role to get the Army, Navy, Air Force, and Marine Corps to experiment together more creatively and develop new weapons and tactics more quickly and economically.[30] But the command's geographic area of responsibility—overseeing U.S. forces in the Atlantic—remained pretty much the same.

Although I am confident that the command will work diligently to accelerate "joint" training and weapons procurement, it is no secret that the decision to rename the Norfolk command was a tactical move by Cohen and his military commanders in order to sidestep a more substantive change that many members of Congress had sought. Two defense experts in the Senate—Dan Coats, a Republican from Indiana, and Joseph I. Lieberman, a Democrat from Connecticut—had drafted legislation to compel the creation of a command devoted exclusively to cross-service innovation.

Because the Joint Forces Command still resides outside the Pentagon's procurement committees and has no statutory authority to enforce its policies or innovations, this newest step toward the Revolution in Military Affairs remains a half-step at best—yet another gesture toward the ideal that fails to hold substance.

How deep should the permanent intermingling of different service elements go? Colonel Douglas Macgregor in *Breaking the Phalanx* argues that the traditional "joint task force" used in recent military operations (in Somalia, Panama, Haiti, and elsewhere) employed a senior Army or Marine Corps general as commander who drew staff from the other services, with their senior officers appointed as deputy commanders (as then Major General H. Norman Schwarzkopf did in Grenada when he was appointed as an "advisor" to the admiral in command). But Macgregor argues—quite forcefully—that this setup is effective only as an *administrative* organization and fails completely in combat. Fighting requires a task organization that mixes the combat units and their support functions in such a way that coordination and mutual support occur naturally.

To that end, Macgregor has proposed what he calls a "joint C^4I battalion" (for command, control, communications, computers, and intelligence) assigned to the Army corps headquarters, Marine expeditionary force, or Navy task force or fleet command in charge of a major military operation. This new unit would operate all of the force's external communications to and from other units and the Pentagon; it would manage various information warfare functions ranging from surveillance using unmanned aerial vehicles to chemical/biological agent detection systems, and would serve as the interface with strategic systems such as the Air Force E-3 AWACS and JSTARS aircraft. *And the unit would be staffed full-time by experts and technicians from all of the services to bring about the "common" battlefield vision twenty-first-century combat will require.*[31]

There are many reasons that shifting from unilateral military service organizations to standing joint forces is impractical and difficult to implement. Bases likely will be closed; tens of thousands of military personnel will have to be transferred from one region to another; and long-standing bonds within the service organizations will be strained. The change will require extensive physical dislocation, expense and effort, and *hard decisions* ranging from selecting site locations to acquiring compatible equipment. For the personnel involved, it will constitute another bout of tumult and stress in a period when military service is already stressful to the breaking point. And it will collide with each military service's most deeply cherished notions of identity and cultural importance.

But the reasons to undertake this task are compelling. First, we can derive greater military efficiency and combat power from joint forces. Second, we will save money in the long run by shedding excess and redundancy. Third, we will help reduce military service parochialism particularly in the competition for procurement and modernization funds.

But the real reason that the U.S. military must restructure itself into a single organization is that it is the only workable way that we can avoid the imminent collapse of the U.S. military.

FIVE STEPS

The Revolution in Military Affairs cannot be won inside the confines of the Pentagon alone. It will require the support of the next presidential administration and Congress, the community of defense analysts and academics who monitor the effort with detached expertise, the news media that reports on military affairs, and the American people—the people who make up the military and those whose tax dollars pay for it.

How can the U.S. military gain that support? First, we must acknowledge that the military as we know it today is in trouble. If we fail to bring about deep reforms we will lose the opportunity to reap the benefits of the information technology revolution. And during the next two decades, as the men and women who make up our military will retire, the force will shrink somewhat—and the tooth-to-tail ratio will grow even worse, from 1:12 to 1:3. (I would not be surprised to see the Army drop from ten divisions today to six; for the 335-ship Navy of today to fall below 190 ships, and for the Air Force to contract from 20 fighter wings today to as few as 12—and without reform there is little we can do about it.) To find solutions, we'll have to transcend our individual loyalties (to the Army, Navy, Air Force, or Marines) and our short-term goals (of getting a particular weapon produced or keeping a certain base open). And *it will take courage* from our President, his secretary of defense, and the Congress. But we have no choice.

Here are my specific proposals for rejuvenating the American Revolution in Military Affairs and carrying the program through to completion:

1. *Launch the debate.* The current state of the U.S. armed forces and the impending collapse of our military capability in the oncoming "defense train wreck" must occupy center stage in the 2000 presidential election and the elections for the 107th Congress. I propose the formation of a coalition of concerned citizens, including retired military leaders, defense experts, and business leaders who can make clear how serious the current military procurement and readiness crisis is and how urgent it is that we bring the military into the next century before its current weapons

become obsolete in 2010 or so. The goals of this information campaign should be

- To inform the American people of the crisis and to muster their support for change;
- To place the subject of the Revolution in Military Affairs and the need for courageous defense reform at the center of the presidential campaign for 2000;
- To draft and publicize a "bipartisan agreement on national defense" to be signed by candidates for President and Congress, affirming that the crisis facing the U.S. military is too serious to become a political football, and that candidates of all parties will agree to discuss and debate the issue with the strict goal of identifying the best solutions for our country's future; and
- To provide an opportunity for candidates for the 107th Congress to affirm that they will refrain from adding amendments to the defense budget only to benefit their particular state or congressional district at the expense of our nation's future defense.

2. *Legislate urgent reforms.* To repeat, the coalition should work to secure within Congress a bipartisan consensus that the impending military situation constitutes a true national emergency and to strive for military reforms unimpeded by pork-barrel concerns or local issues. Legislation to be given immediate consideration includes
   - Passage of "Goldwater-Nichols II" following a set of studies and formal hearings into the state of America's defenses, the need for fulfillment of the Revolution in Military Affairs and defense reform, and specific legislative remedies to mandate and carry out specific reform measures. (See my recommendations for the content of the specific legislation under step 3.)
   - Authorization for two additional rounds of a Base Realignment and Closure Commission under former strict guidelines to avoid political interference from either the Executive Branch or Congress, to bring the military base infrastructure in line with the force reductions since 1986.
   - Hearings and legislation to encourage additional, related reorganization of the Executive Branch to bring the national

security structure of the United States into effective coordination, including reforms of the intelligence community and State Department structures. One pressing need is to modernize the antiquated (nineteenth-century) structure of regional assistant secretaries of state presiding over individual embassies in nations throughout a geographical area, creating a modern design where perhaps a "super-ambassador" who would live in the region could be given overall diplomatic responsibility and authority for the region, mirroring the Pentagon's delegation of authority to a regional commander in chief.

- To air a thorough debate on identifying and safeguarding the country's national security interests in the twenty-first century, and to modernize the national military strategy of the United States to conform with those interests.

3. *Appoint strong Pentagon leadership.* The next President and his administration should affirm its intentions to appoint a secretary of defense committed to carrying out a transformation of the U.S. military to successfully implement true defense reform and the Revolution in Military Affairs. The new secretary should be charged with the responsibility of identifying key issues and changes required to demolish the old system of military service unilateralism and to promote a true joint vision (and budget) for the armed forces. Several immediate steps to be undertaken include
   - *Create a Joint Requirements Committee.* The new administration should seek immediate congressional legislation that will consolidate the identification of military requirements in a Joint Requirements Committee in the Pentagon chaired by the secretary or deputy secretary of defense, with the chairman, Joint Chiefs of Staff, or his designated representative serving as the senior military member and deputy chairman. The membership of the requirements committee should be restricted to the four chiefs of the military services or their vice chiefs, along with four senior civilians from the secretary of defense's staff. The committee would be responsible for setting *all* U.S. military requirements. A combined military-civilian staff would support the committee, constituting the *only* staff dedicated to identifying military requirements in the Defense Department.

The legislation would *eliminate* all authority for defining budget and procurement requirements from the military services, making them responsible to implement the decisions of the Joint Requirements Committee and manage their services. The chiefs of each of the military services would be specifically charged with being the "chief executive officer" of his service's infrastructure, with training and managing personnel and facilities. The secretary of defense's staff would also shift in function, losing all its existing power to set military budget requirements independently. This in turn would justify major reductions in civilian and military Pentagon staff personnel, perhaps by as much as 50 percent of the current manning level. The Joint Requirements Committee would become the center of the Pentagon's primary responsibility to analyze and define the shape of the future U.S. force structure and to determine what military capabilities the force would require.

- *Convene a new "Summit of Key West."* The next President and his national security team immediately upon taking office should convene an "off-site" summit meeting with all of the senior four-star officers of the armed services to outline the goals for the program to implement the Revolution in Military Affairs and real defense reform and to seek guidance and input on specific subjects such as how the Unified Command Plan (which designates the responsibilities of the regional commanders in chief) should be revised.
- *Consolidate key military support functions.* The new secretary of defense should take immediate steps to order the consolidation of the four "great enablers" of combat power—intelligence, communications, logistics, and medical services. An individual military service (Army, Navy, Air Force, or Marine Corps) could be made the executive agent for each of these four joint executive functions, taking on the management responsibilities for all of the four services. In parallel with this consolidation, the defense logistics, communications, and intelligence agencies of each service and the miscellaneous defense agencies should be dismantled and abolished.

- *Transform military education and training.* To set in stage a transformation of U.S. military service culture—in particular, to root out the military service unilateralism and rivalry that have dominated the ranks for over a century—the secretary and his staff should organize for a comprehensive review of all of the military educational institutions, ranging from the military service academies to the National Defense University and armed forces colleges, and to the basic training for enlisted personnel. The impetus of the review should be to identify areas of curriculum and administration reform aimed at radically increasing the exposure and education of military personnel to the full spectrum of U.S. military culture and organization. (For example, the review panel might consider a partial consolidation of the four military service academies to encourage cross-training, mutual indoctrination, and expanded exchange programs.)
- *Establish Standing Joint Force Commands.* Develop a plan for the establishment of four to seven SJFs in each of the five regions under three-star officers, and reporting to the five regional commanders in chief. Disestablish the four-star service component commands that operate as "war fighters" under the CINCs today (and which enforce the single-service mentality around the world). Enormous savings and considerable improvement in military capability would ensue.

We have approached the end of the twentieth century much as historians say we began it. Nearly one hundred years ago we were also flush with military victories and torn by protracted conflicts in obscure and remote lands. We had defeated the Spanish Empire, a military force that pundits of the time proclaimed as "one of the world's most formidable," much as their progeny would describe that of Iraq nine decades later. We had fought in coalition with European nations and the Japanese to enforce "world order" over "the chaos of what is essentially tribal warfare and centuries of animosity" in the Chinese Boxer rebellion.[32] We had begun a long, grinding, and largely unnoticed interdiction effort against the Moros in the southern Philippines to, among other things, "protect the civilized world from the threats of fanaticism and opium."[33] Our predecessors then searched for summary terms, such as "manifest destiny," and spoke of how they had finally expunged the trauma of the Civil War.

We enter a new century facing challenges as daunting as the ones our grandfathers faced in Manila Bay, at Santiago de Cuba, and in the trenches of Belleau Wood, or our fathers at Midway, Normandy Beach, and Okinawa. Our nation, vaulting from the fastness of the continent at the past century's dawn, remains a leader on the world's stage and a power to reckon with for the emerging tyrants of the new era. We remain "the last, best hope" for mankind even in an era where the computer and new information and communication technologies both liberate us from the past while destroying the sense of space and time that for centuries defined our context of existence. As a nation and as a people, we will continue to need an American military force to patrol distant seas, to hold the garrisons of freedom, and to fly missions of war and peace in far-flung regions of the world. It is up to us to decide whether we can safeguard the instruments of our own protection. We must do it *now*.

APPENDIX: ACRONYMS AND EXPLANATIONS FOR TABLE 3-1

Column 1

AWACS: Airborne Warning and Control System installed on the E-3B Sentry aircraft, a modified Boeing 707

Rivet Joint: code name for an electronic intelligence-gathering program using the RC-135 aircraft, also a modified Boeing 707

EP-3A: military designation for a modified P-3 Orion patrol plane used for signals and electronic intelligence-gathering missions

JSTARS: Joint Surveillance and Targeting Radar System, a ground-watching radar installed on the E-8 aircraft, another modified Boeing 707, which can track moving vehicles from several hundred miles away; first used in the 1991 Gulf War

HASA: High-Latitude Surveillance Architecture, a description for a number of aircraft- and satellite-based surveillance systems

SBIRS: Space-Based Infrared System program, a proposed galaxy of satellites planned for initial launch in 2002 that will constitute a single "system of systems" supporting missile warning, missile defense, and intelligence applications

Tier 2/Tier 3: a proposed theater ballistic-missile defense program utilizing the Navy's Aegis missile cruisers firing Tier 2 (medium-altitude) and Tier 3 (high-altitude) missiles

TARPS: Tactical Aerial Reconnaissance Photo System, a pod mounted on F-14s or other tactical aircraft that can take low- and medium-altitude imagery of a target area, used mainly for pre-strike and post-attack battle damage assessment

MTI: Moving Target Indicator

UGS: Unmanned Ground Sensor, used in wide-area surveillance operations including tracking enemy ground forces or monitoring for underground nuclear detonations

ISAR: Inverted Synthetic Aperture Radar

Magic Lantern: a helicopter-installed blue laser used for detecting antiship mines in shallow water

Predator: an unmanned aerial vehicle used for reconnaissance

NVG: night-vision goggles

Cobra Ball: an aerial surveillance platform using RC-135 reconnaissance aircraft to study foreign ballistic missile launches

Column 2

GCCS: Global Command and Control System, the Pentagon's new worldwide communications network for controlling U.S. forces

MILSTAR: a new communications satellite originally designed as a backup system that could survive the effects of a nuclear war

JSIPS: Joint Services Imagery Processing System, a communications network that allows digital images from all services to be delivered to users

DISN: Defense Information Services Network, a term for the entire U.S. military communications system

SABER: a term for "identification, friend or foe" technology that allows the U.S. military to identify aircraft, ships, or ground vehicles as friendly through the use of coded transponders that emit a signal upon activation by the U.S. aircraft

C^4IFW: a compound acronym, with "C^4I" standing for command, control, communications, computers, and intelligence, and "FW" standing for "for the warrior," referring to a program to disseminate situational awareness and vital intelligence data to the lowest echelons of combat soldiers, sailors, airmen, and Marines on the battlefield

TADIL J: Tactical Data Information Link (J), an aircraft-to-aircraft secure communications channel

TACSAT: for Tactical Satellite communications

TRAP: a code word for the broadcast of a sanitized version of images captured by highly classified reconnaissance satellites

MIDS: a version of the U.S. military Link 16 tactical data communications system approved for export to other nations

SONET: a commercial routing and switching system for a complex communications network

Column 3

SFW: Sensor Fused Weapons, smart munitions

JDAM: Joint Direct Attack Munitions, a precision-guided weapon kit attached to a 1,000-pound or 2,000-pound gravity bomb, maneuvered to a pinpoint location on the ground by the Global Positioning System navigational satellites; first used in Kosovo.

JSOW: Joint Standoff Weapon, a proposed Navy and Air Force air-to-surface guided missile that glides to its target using Global Positioning System inputs for precise accuracy

TLAM: Tomahawk Land Attack Missile, a designation for the Tomahawk ship-launched cruise missile

ATACMS/BAT: Army Tactical Ballistic Missile System/Brilliant Antitank warhead, a surface-to-surface conventionally armed missile first used in the 1991 Gulf War

CALCM: Conventional Air-Launched Cruise Missile, a conventionally armed cruise missile launched from bomber aircraft; used in both the Gulf War and Kosovo

HARM: High-speed Anti-Radiation Missile, an air-to-ground missile that homes in on enemy radar emissions to destroy the radar

THAAD: Theater High-Altitude Air Defense, a proposed Army theater antiballistic missile system to shoot down enemy missiles

Hellfire II: an advanced version of the Hellfire antitank missile fired by Army attack helicopters

Javelin: a shoulder-fired antitank missile

Longbow: a surveillance and targeting system installed on new versions of the AH-64 Apache attack helicopter

LOSAT: a mortar system that fires precision-guided shells

HAVE NAP: a medium-range air-to-ground missile launched from heavy bombers, originally developed by Israel

AGM-130: a military term for an air-to-surface missile kit attached to a 2,000-pound gravity bomb that enables the attack aircraft to guide it to the target via a TV image transmitted from the weapon to the aircraft cockpit

SUGGESTED READING

BOOKS

Adams, James. *The Next World War—Computers Are the Weapons and the Front Line Is Everywhere.* New York: Simon and Schuster, 1998.

Atkinson, Rick. *Crusade—The Untold Story of the Persian Gulf War.* New York: Houghton Mifflin, 1993.

Beckwith, Colonel Charlie, with Donald Knox. *Delta Force.* New York: Dell, 1983.

Bowden, Mark. *Black Hawk Down—A Story of Modern War.* New York: Atlantic Monthly Press, 1999.

Clausen, Henry C., and Bruce Lee. *Pearl Harbor: The Final Judgment.* New York: Crown Publishers, 1992.

Dunnigan, James F. *Digital Soldiers.* New York: St. Martin's Press, 1996.

Friedman, George and Meredith. *The Future of War.* New York: Crown Publishers, 1996.

Gabriel, Richard A. *Military Incompetence: Why the U.S. Military Doesn't Win.* New York: Hill and Wang, 1985.

Gannon, Michael. *Operation Drumbeat.* New York: Harper and Row, 1990.

Gouré, Daniel and Jeffrey Rammey. *Averting the Defense Train Wreck in the New Millennium.* Washington, D.C.: CSIS Press, 1999.

Gordon, Michael, and Lieutenant General Bernard E. Trainor. *The Generals' War—The Inside Story of the Persian Gulf War.* New York: Little, Brown and Company, 1995.

Greider, William. *Fortress America.* New York: PublicAffairs, 1998.

Hart, Gary. *The Minutemen—Restoring an Army of the People.* New York: The Free Press, 1998.

Hart, Gary, with William S. Lind. *America Can Win—The Case for Military Reform.* Bethesda, Md.: Adler and Adler, 1986.

Kaplan, Robert D. *The Ends of the Earth.* New York: Vintage Departures, 1997.

Macgregor, Lieutenant Colonel Douglas A. *Breaking the Phalanx—A New Design for Landpower in the Twenty-first Century.* Westport, Conn.: Praeger; published in cooperation with the Center for Strategic and International Studies, 1997.

Odom, William E. *America's Military Revolution.* Washington, D.C.: The American University Press, 1993.

Owens, Admiral William A. *High Seas.* Annapolis, Md.: U.S. Institute Press, 1995.

Perry, Mark. *Four Stars.* Boston: Houghton Mifflin, 1989.

Polmar, Norman, and Thomas B. Allen. *World War II: America at War.* New York: Random House, 1991.

Regan, Geoffrey. *Blue on Blue: A History of Friendly Fire.* New York: Avon Books, 1995.

Scales, Brigadier General Robert H., and the Desert Storm Special Study Group. *Certain Victory: The U.S. Army in the Gulf War.* Office of the Chief of Staff, United States Army, 1993.

Schwartu, Winn. *Information Warfare—Chaos on the Information Superhighway.* New York: Thunder's Mouth Press, 1994.

Schwartzstein, J. D., editor. *The Information Revolution and National Security.* Washington, D.C.: Center for Strategic and International Studies, 1996.

Schwarzkopf, H. Norman, with Peter Petre. *It Doesn't Take a Hero.* New York: Bantam Books, 1992.

Shukman, David. *Tomorrow's War—The Threat of High-Technology Weapons.* Orlando, Fla.: Harcourt Brace and Company, 1995.

Taylor, Maxwell D. *The Uncertain Trumpet.* New York: Harper and Brothers, 1959.

Van Creveld, Martin. *The Transformation of War.* New York: The Free Press, 1991.

REPORTS AND OTHER PUBLICATIONS

America's Overseas Presence in the Twenty-first Century. The Report of the Overseas Presence Advisory Board, U.S. Department of State, November 1999.

Annual Report to the White House and the Congress. Message from Secretary of Defense William A. Cohen. May 1998. Downloaded from the Department of Defense website at *www.defenselink.mil.*

Army Focus 1994: Force XXI—America's Army in the Twenty-first Century. Washington, D.C.: Department of the Army, 1999.

Epley, William W. *Roles and Missions of the United States Army: Basic Documents with Annotations and Bibliography.* Washington D.C.: U.S. Army Center of Military History, 1991.

Equipped for the Future—Managing U.S. Foreign Affairs in the Twenty-first Century. The Project on the Advocacy of U.S. Interests Abroad. Schall, John, executive director, by the Henry L. Stimson Center. Washington, D.C., October 1998.

Final Report of the Division XXI Advanced Warfighting Experiment. Fort Leavenworth, Kans.: U.S. Army Training and Doctrine Command Analysis Center, July 1998.

Hard Choices—Fighter Procurement in the Next Century. A policy analysis paper by Williamson Murray given at the Cato Institute, Washington, D.C., February 26, 1999.

Joint Vision 2010, Office of the Chairman, Joint Chiefs of Staff, 1995, and *Concept for Future Joint Operations—Expanding Joint Vision 2010,* Office of the Chairman, Joint Chiefs of Staff, Washington, D.C.: Pentagon, 1997.

New World Coming: American Security in the Twenty-first Century—Major Themes and Implications. The United States Commission on National Security/Twenty-first Century, September 15, 1999.

Owens, William and Joseph Nye. "America's Information Edge." *Foreign Affairs.* April 1996.

Reinventing Diplomacy in the Information Age. Burt, Richard and Robison, Olin, project co-chairs, Center for Strategic and International Studies, Washington, D.C., October 9, 1998.

Report of the Select Committee on U.S. National Security and Military/Commercial Concerns with the People's Republic of China. Washington, D.C.: Government Printing Office, 1999.

Soviet Military Power. Department of Defense, 1989 and 1990 editions.

Strategic Assessment 1998—Engaging Power for Peace. Institute for National Strategic Studies. Washington, D.C.: Government Printing Office, 1998.

"The FY2000 Defense Budget: Gambling with America's Defense." By Representative Floyd Spence. *National Security Report,* 3, no. 1 (February 1999).

The Quadrennial Defense Review. May 1997. Downloaded from the Department of Defense website at *www.defenselink.mil.*

"U.S. Space Almanac." *Air Force* (magazine). August 1999, p. 39.

Who Goes There: Friend or Foe? A report by the Office of Technology Assessment, June 1993.

NOTES

INTRODUCTION

1. *The Defense Monitor* 27, no. 1, Center for Defense Information, 1999.
2. FactSheet on the 2000 Pentagon budget, GOP Conference Staff, April 22, 1999.
3. "The Coming Defense Trainwreck," Center for Strategic and International Studies analysis project information report, March 23, 1999.
4. Ibid.
5. Gary Hart, *The Minutemen—Restoring an Army of the People* (New York: The Free Press, 1998), p. 154.
6. Dominic A. Paolucci, quoted in *Rickover*, by Norman Polmar and Thomas B. Allen (New York: Simon and Schuster, 1982), p. 421.
7. General Frederick M. Franks, commander of the Army VII Corps in Operation Desert Storm, in a 1994 address to the Association of the U.S. Army in Orlando, Florida.
8. Sun Tzu, *The Art of War*, trans. Ralph D. Sawyer (Boulder: Westview Press, 1994), p. 135.
9. *The Military Maxims of Napoleon, in Roots of Strategy*, pp. 440–41.
10. Antoine-Henri Jomini, *The Art of War*, trans. G. H. Mendell and W. P. Craighill (Westport, Conn.: Greenwood Press, 1971), p. 63.
11. Ibid.
12. Carl von Clausewitz, *On War*, ed. and trans. by Michael Howard and Peter Paret (Princeton: Princeton University Press, 1984), p. 97.
13. Ibid., p. 101.
14. Niccolò Machiavelli, *The Prince*, in *The Portable Machiavelli*, trans. and ed. by Peter Bondanella and Mark Musa (New York: Penguin, 1979), p. 94.

15. Within a week of the launching of U.S.–led air strikes against Serbian targets in Kosovo and Serbia last March, the Associated Press, quoting U.S. intelligence officials, reported that Serbian defense officials had traveled to Baghdad a month before the conflict erupted to confer with their Iraqi counterparts, presumably to determine how the Iraqis have adjusted to continuous American aircraft and cruise missile strikes. "Yugoslavia, Iraq Talked Air-Defense Strategy," by John Diamond, the Associated Press, March 30, 1999.
16. Vlahos, Michael, "The War After Byte City," in *The Information Revolution and National Security*, ed. by Stuart J. D. Schwartzstein (Washington, D.C.: The Center for Strategic and International Studies, 1996), p. 86.
17. Ibid., p. 88.

CHAPTER 1

1. The multiservice unified commands are responsible for conducting major combat operations in specified geographical areas or other specific duties as directed by the President and the secretary of defense, with orders from the civilian leadership passing through the chairman, Joint Chiefs of Staff. They are the U.S. Atlantic Command, Norfolk, Va.; U.S. Central Command, MacDill Air Force Base, Florida; U.S. European Command, Stuttgart-Vaihingen, Germany; U.S. Southern Command, Miami, Florida; U.S. Pacific Command, Camp H. M. Smith, Hawaii; U.S. Space Command, Peterson Air Force Base, Colorado; U.S. Special Operations Command, MacDill Air Force Base, Florida; U.S. Strategic Command, Offutt Air Force Base, Nebraska; and U.S. Transportation Command, Scott Air Force Base, Illinois.
2. A major Pentagon assessment of defense requirements in 1997 concluded, "Absent a marked deterioration in world events, the nation is unlikely to support significantly greater resources dedicated to national defense than it does now—about $250 billion in constant 1997 dollars per year." So far that estimate has held firm. *The Quadrennial Defense Review*, Section 9, May 1997, downloaded from *www.defenselink.mil* on May 4, 1999. Also, during an appearance before the House Armed Services Committee on February 24, 1999, to discuss the proposed fiscal year 2000 defense budget, the chiefs of the armed services warned that at best the $267.2 billion budget plan would only lock in the status quo for another year. It would not reverse the slide in readiness, the increasing age of weapons, or the declining condition of family housing units and barracks, officials said. See "Chiefs: Clinton Plan Won't Repair Readiness," by William Matthews, *Navy Times*, March 8, 1999.
3. *New World Coming: American Security in the Twenty-first Century—Major Themes and Implications*, The United States Commission on National Security/Twenty-first Century, September 15, 1999.
4. *The Quadrennial Defense Review*, May 1997, downloaded from the Department of Defense website at *www.defenselink.mil.*

5. James Adams, *The Next World War* (New York: Simon and Schuster, 1998), p. 279.
6. The issue of higher-than-anticipated costs and the escalation of demands by regional commanders for more and more units in crises is addressed in detail in Chapter 2. The most recent example occurred last year when, less than three weeks after NATO forces unleashed air strikes against Serbian forces in Yugoslavia on March 24, 1999, the Pentagon Joint Staff more than doubled the cost estimate for Operation Allied Force, from $2 billion cited by Undersecretary of Defense John Hamre in early March to more than $5.8 billion covering the period March 24, 1999, through September 2000. The changed estimate was first revealed by the newsletter *Inside the Air Force* in its April 9, 1999 issue. Also, the Bosnian peacekeeping operation by April 1999 had exceeded $10 billion in costs, according to Pentagon officials. Bill Gertz and Rowan Scarborough, "Stretched Thin," *The Washington Times*, April 9, 1999.
7. Bob Woodward, *The Commanders* (New York: Pocket Star Books, 1991), p. 82.
8. Powell in his memoirs defended the Base Force concept as "a realistic military posture for a future in which two superpowers would not be flexing their muscles at each other." He described the goal of the 1990 plan as "to build a leaner, more efficient, high quality force capable of any mission." See Colin L. Powell, *My American Journey* (New York: Random House, 1995), pp. 546 and 551.
9. Edward Luttwak, *The Pentagon and the Art of War* (New York: Institute for Contemporary Studies/Simon and Schuster, 1984), p. 18.
10. As senior military assistant to Secretary of Defense Dick Cheney in 1989–90, I observed the compromises and discussions that led to the creation of the Base Force concept firsthand.
11. Maxwell D. Taylor, *The Uncertain Trumpet* (New York: Harper and Brothers Publishers, 1959), p. xi.
12. Don M. Snider, Daniel Goure, and Stephen A. Cambone, "Defense in the Late 1990s—Avoiding the Train Wreck," a memorandum generated by the Center for Strategic and International Studies, Washington, D.C., March 1999.
13. Gertz and Scarborough, "Stretched Thin," p. 8.
14. Rowan Scarborough, "Air Force Search-and-Rescue Operations Called 'Broken,' " *The Washington Times*, September 13, 1999.
15. Representative Floyd Spence, "The FY2000 Defense Budget: Gambling with America's Defense," *National Security Report*, February 1999.
16. Lane Pierrot and Jo Ann Vines, "A Look at Tomorrow's Tactical Air Forces," Congressional Budget Office, Washington, D.C., 1997.
17. Williamson Murray, "Hard Choices—Fighter Procurement in the Next Century," a policy analysis paper, Cato Institute, February 26, 1999, p. 11.
18. Ibid.
19. *Quadrennial Defense Review*, May 1997, Chapter VII.
20. Strictly speaking, U.S. military operations in the Philippines during World War II took place in the Commonwealth of the Philippines, a guardianship of the United States, rather than in the United States per se. The World War II operations in the

Aleutian Islands against the Japanese came closer to a campaign on U.S. soil, but at the time, the Aleutians, being part of the then U.S. Alaska Territory, did not have statehood.

21. "Military Strength Figures for Jan. 31, 1999 Summarized by the Department of Defense," news release, March 8, 1999.
22. Ibid.
23. Pentagon Youth Attitude Tracking Study.
24. Steven Komarow, "Army Offers $20,000 to Sign Up," *USA Today*, November 19, 1999, p. 1.
25. Several works emerged in the mid-1950s that had a salient impact both inside the military and in the broader academic/journalist community regarding the implications of a large, professional military officer corps. Samuel P. Huntington's *The Soldier and the State* (Cambridge, Mass.: Harvard University Press, 1957); Morris Janowitz's *The Professional Soldier* (New York: The Free Press, 1960); and Alfred Vagts' *A History of Militarism: Civilian and Military* (New York: Meridian Books, 1959) were among the most important. There was a flurry of renewed interest in the 1970s, driven in part by the American experience in Vietnam, and the shift to an all-volunteer military. See, for example, Adam Yarmolinsky, *The Military Establishment: Its Impacts on American Society* (New York: Harper and Row, 1971); and the work of the Inter-University Seminar on Armed Forces, particularly, Charles C. Moskos, Jr., ed., *Public Opinion and the Military Establishment* (Beverly Hills, Calif.: Sage Publications, 1971).
26. Perry M. Smith, "The Military and American Society," in Richard G. Head and Ervin J. Rokke, eds., *American Defense Policy* (Baltimore, Md.: The Johns Hopkins University Press, 1973), p. 500.
27. Investigative journalist Seymour Hersh concluded in his study of "Gulf War Syndrome" that the Clinton Administration's uneasy relationship with the military had left the White House disinclined to press the Pentagon on searching for answers to the growing number of sick Gulf War veterans. See *Against All Enemies: Gulf War Syndrome—the War Between America's Ailing Veterans and Their Government*, by Seymour Hersh (Library of Contemporary Thought, 1998).
28. The Aegis combat system is built around a very powerful phased-array radar that allows the ship to detect, identify, and track numerous aircraft at great distances, computer systems that provide very good targeting information, and air defense missiles that have long range and high accuracy.
29. Budget estimates and the analytic machinations to convert them to a common dollar base are notoriously inaccurate. Like counts of men or weapons systems, budget figures can only hint at relative military capabilities. But the amounts different nations spend on their militaries can provide a gross sense of how each sees its military and how the military fits into the nation's priorities. By themselves, budget figures can only hint at such things. Expressed in terms of the portion of all national spending, or on a per capita basis, however, budget figures can add a bit more insight.

30. Tables from the Secretary of Defense's 1997 *Annual Report to the President and the Congress* provide an overview of today's force. I've used these tables for a couple of reasons. First, they present the canonical description of U.S. forces. Although the tables are taken from an official document, the description they provide is echoed in the work of many journalists, academics, and other observers. Second, the tables tell us a lot about how the Defense Department and many others think about our forces. The categories used in the tables are a guide to current assumptions about the interdependencies—or lack of them—among the various components that make up today's force.
31. Ibid.
32. This is not to say the Army and Marines have been totally consistent in this approach throughout the twentieth century. During World War I the Marine Corps served as part of the American Expeditionary Force under Army General John Pershing, and six decades later in Vietnam occupied, patrolled, and fought in a region of South Vietnam for several years. And the Army conducted major amphibious landings in World War II in North Africa, Sicily, and France as well as in the South Pacific and Philippines. But in the post-Vietnam era the evolution of Army and Marine Corps practice has been to focus on the specific approaches cited in the chapter.
33. *Annual Report to the President and the Congress for the 1999 Fiscal Year*, message from Secretary of Defense William Cohen, p. 1.
34. An example of the military system's flaccid response to the rapid-paced information age was the planned refueling of the nuclear-powered carrier USS *Enterprise* during 1993–95. When the carrier entered Newport News Shipbuilding, the first renovation was a total replacement of the carrier's Combat Direction Center prior to the nuclear refueling. Less than thirty months later, upon completion of the overhaul of its eight reactors, the *Enterprise* returned to another part of the shipyard where it—once again—had its Combat Direction Center stripped out and another array of computers installed. Source: confidential Navy official.

CHAPTER 2

1. Originally named the *Tbilsi* (for the capital of the Soviet Republic of Georgia), the *Admiral Kuznetsov* had held the U.S. Navy's full attention since the early 1980s when its construction began. The 65,000-ton aircraft carrier was only 100 feet shorter in length than the U.S. Navy's powerful *Nimitz*-class aircraft carriers, and had already demonstrated an ability to launch and recover modern combat aircraft, including modified versions of the Su-27 Flanker and MiG-29 Fulcrum fighters, and the Su-25 Frogfoot strike bomber. At the time of its appearance in the Mediterranean in 1991, the Soviets were believed to be building a second carrier of this type, the *Varyag*, and the lead ship of a larger, follow-on aircraft carrier design was under construction. Although the *Admiral Kuznetsov* carried a smaller number of

aircraft than its U.S. counterparts, the Soviet carrier was also designed with launching tubes to fire 12 SS-N-19 antiship missiles with nuclear warheads.

2. The strategy of the Soviet fleet included missions to operate and protect its missile submarines, protect Soviet coastal waters from hostile attack, and support Soviet land forces by providing naval fire and logistical support, according to the report *Soviet Military Power 1990*, Department of Defense, p. 82.
3. An excellent essay on the historic nature of war and the levels of violence can be found in *Brassey's Encyclopedia of Military History and Biography* (Washington, D.C.: Brassey's, 1994), pp. 1045–64.
4. Douglas A. Macgregor, Lieutenant Colonel, USA, *Breaking the Phalanx—A New Design for Landpower in the Twenty-first century* (Westport, Conn.: Praeger, in cooperation with the Center for Strategic and International Studies, 1997), p. 31.
5. Martin van Creveld, *Technology and War: From 2000 B.C. to the Present* (New York City: The Free Press, 1989).
6. John Keegan, *A History of Warfare* (New York: Vintage Books, 1993), pp. 200–205.
7. Bevin Alexander, *How Great Generals Win* (New York: Avon Books, 1993), p. 73.
8. Alexander, *How Great Generals Win*, p. 79.
9. James Adams, *The Next World War—Computers Are the Weapons and the Front Line Is Everywhere* (New York: Simon and Schuster, 1998), p. 56.
10. Keegan, *A History of Warfare*, p. 355.
11. Macgregor, *Breaking the Phalanx*, p. 38.
12. In 1999 the U.S. Army maintained four active corps headquarters: I Corps at Fort Lewis, Washington, responsible for operations in the Asia-Pacific region as well as training responsibilities for reserve and National Guard units nationwide; III Corps at Fort Hood, Texas, supervising a number of "heavy" divisions and responsible for implementing the service's Force XXI digital technology; XVIII Airborne Corps at Fort Bragg, North Carolina, the nation's "911 service" that controls the 82nd Airborne Division, 101st Air Assault Division, 3rd Infantry Division, and 10th Mountain Division; and the V Corps headquartered in Germany and responsible for U.S. Army units in Europe including those designated for peacekeeping operations in Bosnia and Kosovo. Each corps headquarters, led by a three-star lieutenant general, is structured to control combat operations of between two and five U.S. Army divisions.
13. *Official Records of the War Against the Rebellion*, XXXII, Part 3, p. 312, cited in Geoffrey Perret, *Ulysses S. Grant, Soldier and President* (New York: Random House, 1997), p. 301.
14. During the four years of World War I, approximately 10 million soldiers from all the warring states in Europe were killed. During World War II, which began in Asia in 1937 and in Europe two years later, and ran until the surrenders of Nazi Germany and imperial Japan in May and September 1945, respectively, more than 22 million military and civilian personnel were killed. This includes 408,000 American casualties including 294,597 combat deaths. Other military fatalities included 7.5 million Soviet troops; 3.25 million Germans; 1.5 million Japanese;

544,000 British Commonwealth troops; and 300,000 Italians. Source: Norman Polmar and Thomas Allen, *World War II: America at War 1941–45* (New York: Random House, 1991), p. 193.

15. Although formed by the combination of the two German words *blitz*, for "lightning," and *krieg*, for "war," the term is actually not a German word. Tradition has it that a British newspaper editor coined the word in 1940. See *Brassey's Encyclopedia of Military History and Biography*, p. 161.
16. A saga unto itself, the U.S. military's rebirth after Vietnam is best told in *Prodigal Soldiers—How the Generation of Officers Born of Vietnam Revolutionized the American Style of War*, by James Kitfield (New York: Simon and Schuster, 1995).
17. William J. Perry, "Desert Storm and Deterrence," *Foreign Affairs* 70 (Fall 1991): 66–82.
18. Hints that the USSR might not be the growing military power portrayed by the Weinberger administration at the time were not something that was easy to argue inside the Pentagon in the mid-1980s. The hypothesis was suggested under tight security wraps. It surfaced publicly, however, in 1988 as part of the work of the Pentagon's Commission on Integrated Long-Term Strategy. See *The Future Security Environment: Report of the Future Security Environment Working Group* (Washington, D.C.: Government Printing Office, October 1988).
19. "Kosovo Will Be to Post-2000 Defense As Gulf War to 1990s," *Defense Daily*, April 27, 1999, p. 7.
20. Rick Atkinson, *Crusade—The Untold Story of the Persian Gulf War* (Boston and New York: Houghton Mifflin, 1993, p. 2.
21. See James F. Dunnigan and Raymond M. Macedonia, *Getting It Right: American Military Reforms After Vietnam to the Gulf War and Beyond* (New York: William Morrow and Company, 1993), p. 211. Citing historical accounts, the authors calculated that the U.S. Army in Operation Desert Storm covered 368 kilometers in four days for an average movement of 92 kilometers per day. This exceeded the 1945 Soviet invasion of Manchuria (82 kilometers per day), the British invasion of Sinai against Turkey in 1918 (56 kilometers per day), the Israeli sweep across the same Sinai Peninsula against Egypt in 1967 (55 kilometers per day), and the Soviet invasion of eastern Germany in 1944 (50 kilometers per day).
22. The most graphic contrast in the authority of mission commanders to control the operations and organize them based on mission requirements—instead of responding to political pressure—can be seen in the 1980 Iran rescue mission and Operation Desert Storm. During the planning process for the embassy hostage rescue, the Marine Corps insisted that some of its personnel be allowed to take part in the raid, since all of the other combat services had a role. As a result, poorly prepared Marine helicopter pilots were given the role of flying the massive MH-53 helicopters from the USS *Nimitz* to the initial landing site, replacing Air Force pilots already highly trained for long-distance night flying. Thus, an insufficient number of helicopters made it to the rendezvous and the mission was aborted—only to descend into complete failure when one of the helicopters crashed into a refueling aircraft on the desert airstrip. Eleven years

later, Schwarzkopf, as overall commander in chief of Operation Desert Storm, canceled a planned Marine amphibious landing in Kuwait in favor of a deception operation that led the Iraqis to *think* an invasion was inevitable. Despite immense political pressure from Marine Corps Commandant General Al Gray and his subordinates to proceed with the landing of 17,000 Marines embarked on amphibious ships in the Persian Gulf, which would have been the biggest Marine assault since Inchon, South Korea, in 1950, Schwarzkopf said no, and the decision stood—thanks to the Goldwater-Nichols Defense Reform Act. (See Atkinson, *Crusade*, pp. 169–72 and 237–40.

23. The commanders under General H. Norman Sckwarzkopf included Lieutenant General Calvin A. H. Waller, deputy commander of Central Command; Army Lieutenant General John Yeosock, Army forces commander; Lieutenant General Fred Franks and Lieutenant General Gary Luck, commanders of the VII Corps and XVIII Airborne Corps, respectively; Lieutenant General Charles Horner, commander of Central Command air forces; Vice Admiral Stan Arthur, commander of Central Command naval forces; and Lieutenant General Walt Boomer, commander of Central Command Marine Corps forces. Schwarzkopf's equal-in-name partner was Saudi Prince General Khaled Bin Sultan, who commanded all allied forces.
24. H. Norman Schwarzkopf, *It Doesn't Take a Hero* (New York: Bantam Books, 1992), p. 252.
25. Jeffrey R. Cooper, "Strategy," in *Air and Space Power in the New Millennium*, ed. by Daniel Goure and Christopher M. Szara (Washington, D.C: Center for Strategic and International Studies, 1997), p. 51.

CHAPTER 3

1. James Adams, *The Next World War* (New York: Simon and Schuster, 1998), pp. 14–15.
2. The descriptions of D-Day events and issues are drawn from *World War II—America at War*, by Norman Polmar and Thomas B. Allen (New York: Random House, 1991); *There's a War to Be Won—The United States Army in World War II*, by Geoffrey Perret (New York: Random House, 1991); and *D-Day*, by Stephen E. Ambrose (New York: Simon and Schuster, 1994).
3. Ambrose, *D-Day*, p. 222.
4. Ibid., p. 295.
5. Omar Bradley and Clay Blair, *A General's Life: An Autobiography* (New York: Simon and Schuster, 1983), pp. 249–51.
6. Ambrose, *D-Day*, p. 321.
7. Descriptions of the Operation Desert Storm ground campaign are drawn from *Certain Victory: The U.S. Army in the Gulf War*, by Brigadier General Robert H. Scales and the Desert Storm Special Study Group (Office of the Chief of Staff, United

States Army, 1993); *Crusade—The Untold Story of the Persian Gulf War*, by Rick Atkinson (Boston: Houghton Mifflin, 1993); *Airpower in the Gulf*, by James P. Coyne (Arlington, Va.: Airforce Association, 1992); *Storm over Iraq—Air Power and the Gulf War*, by Richard P. Hallion (Washington, D.C. & London: Smithsonian Institution Press, 1992); and *The Generals' War*, by Michael Gordon and Bernard Trainor (New York: Little, Brown, 1995).

8. Scales, *Certain Victory*, p. 59.
9. Ibid., p. 41.
10. Ambrose, *D-Day*, p. 576; Coyne, *Airpower in the Gulf*, p. 102.
11. William J. Perry, "Desert Storm and Deterrence," *Foreign Affairs*, April 1995, pp. 67–82.
12. This section draws on Scales's *Certain Victory*, Atkinson's *Crusade*, and Hallion's *Storm over Iraq*.
13. Hallion, *Storm over Iraq*, p. 188.
14. Scales, *Certain Victory*, pp. 371–72.
15. Memorandum from Brigadier General Steve Arnold, cited in Atkinson, *Crusade*, p. 339.
16. Gordon and Trainor disclosed that on the eve of the ground offensive, then CIA Director William Webster provided the White House with the CIA's bomb damage assessment, which concluded that only 524 Iraqi tanks (12 percent) had been destroyed in the six-week air campaign, contrasted with the Central Command estimate of 1,688 tanks (39 percent of the Iraqi force). "Webster's assessment had enormous implications. . . . If the American military had badly underestimated the damage done to the Iraqis, it could suffer considerable casualties when it attacked," the authors noted. The Defense Department gave little weight to the CIA analysis and in retrospect Central Command had the more accurate estimate. (Gordon and Trainor, *The General's War*, p. 335.) Also see Scales, *Certain Victory*, p. 187.
17. General Colin Powell, press conference, January 23, 1991.
18. As a top secret U.S. commando operation in early 1970 trained at a secret Florida site preparing to rescue American prisoners of war at the Son Tay POW camp, concern that Soviet low-orbit reconnaissance satellites would discover and compromise the plan prompted the military commander to order a 2 × 4 lumber and canvas mock-up of the North Vietnamese prison to be disassembled and hidden each day when the Soviet satellite appeared overhead. See *The Raid*, by Benjamin F. Schemmer (New York: Avon, 1976), p. 92.
19. Radar satellites are believed to have the potential to detect submerged submarines by tracking minute traces of the submarine's underwater propeller wake out of the random pattern of surface water movement. Details of the phenomenon are highly classified, but evidence of U.S. interest emerged last year when news reports alleged a Chinese national had delivered information on research into radar detection of submarines to Chinese intelligence officials. See "Reports a Scientist Gave U.S. Radar Secrets to China," *The New York Times*, May 10, 1999, p. A-1.

20. See "Doomed Satellite," *USA Today*, May 14, 1999, p. A-3.
21. *Annual Report*, May 1998, Chapter 7, p. 1.
22. Ibid.
23. Vipin Gupta, "New Satellite Images for Sale," *International Security* 20, no. 1 (Summer 1995): 94–125.
24. Some analysts have begun using the term *high vista* to describe the latent ability of U.S. space-based sensors to track and identify significant objects and events at distances greater than previous systems could perform—because the sensors through which that object or event is "seen" can draw from a wider range of emissions of electromagnetic energy.
25. The Treaty on Intermediate-range Nuclear Forces banned ground-launched cruise missiles with ranges greater than 800 kilometers. The treaty defined ground-launched cruise missiles in engineering terms in such a way that banned UAVs capable of flying 800 kilometers that carry weapons of any type.
26. As far back as the late 1950s, the United States had deployed massive arrays of underwater acoustic sensors to help track the movement of Soviet submarines. When the submarine USS *Scorpion* disappeared in the eastern Atlantic in May 1968, subsequent investigation discovered recordings of its sinking from sites in the Canary Islands and Argentia, Canada, that helped searchers pinpoint the location of the wreckage 400 miles southwest of the Azores. See "The USS *Scorpion*: Mystery of the Deep," by Ed Offley, *The Seattle Post-Intelligencer*, May 21, 1998 (located at http://www.seattle-pi.com/awards/scorpion/index.html).
27. Because the longer-wavelength SARs can penetrate foliage, we are focusing a great deal of effort on improving automatic target recognition techniques and processing for SARs operating at the UHF and lower frequencies.
28. Until recently, SAR imagery was formed by applying fast Fornier transforms to raw radar signals.
29. "Yugoslav Air War's Cost to U.S. May Top $4 Billion, Experts Say," by Tom Raum, *The Philadelphia Inquirer*, June 29, 1999.
30. See Martin Libicki, *The Intersystem* (Washington, D.C.: National Defense University Press, forthcoming).
31. There are some differences built into the satellite transmitters. The satellites transmit two types of code signals, a CA code for the public that allows the receiver to locate itself within 100 meters of its actual location, and a Y code that allows the military user to estimate his actual location within about 18 meters. But it is possible to achieve greater accuracy through various techniques, among which is one called differential GPS. Here, the basic approach is to measure the errors in the signals received from each GPS satellite in view at a precisely known location, and then to broadcast these errors to the user receiver. This allows a positioning accuracy within centimeters of the receiver. Anything carrying the receiver—individual soldier, aircraft, land vehicle, ship, or missile—can therefore know—precisely—where he or it is at any time two or more of the GPS satellites are within sight. (And because a constellation of satellites is involved, virtually every point on or

above the earth is always within sight of two or more satellites.) This remains the case as long as the signal from the satellites is not jammed or broken by vegetation or other intervening factors.

32. Jamming is particularly effective in blocking receipt of the CA-coded transmissions used by nonmilitary GPS receivers. A 1-watt jammer less than about 22 kilometers away, under optimal conditions, can block receipt of a CA-coded signal by a civil receiver. Because the Y-coded signals the military uses are broadcast using spread spectrum and most military GPS receivers have sophisticated antenna and signal nulling devices, it is significantly more difficult—on the order of a million times—to jam the GPS signals the military uses. Jamming the CA signal, however, makes it harder for the military to use the CA signal to calibrate its own timing to decrypt the Y signal.
33. The Joint Direct Attack Munitions, or JDAM, is essentially a GPS receiver and guidance kit that, when attached to a glide bomb, provides an attack standoff range of up to 20 kilometers. Recent JDAM costs ran about $18,000 per kit, and prototypes suggest that shorter-range munitions can be retrofit for several thousand dollars. In contrast, a weapon carrying its own seeker normally runs from about $100,000 to $1 million.

CHAPTER 4

1. See Keane, "Jointness Must Be Taken Seriously or Services May Lose Control," *Inside the Army*, May 17, 1999, p. 1.
2. See Henry C. Clausen and Bruce Lee, *Pearl Harbor: The Final Judgment* (New York: Crown Publishers, 1992), particularly pp. 229–45. Clausen, an Army staff lawyer in 1944, was assigned by then Secretary of War Henry Stimson to conduct a top secret analysis of the Pearl Harbor attack and was given extraordinary access to senior military commanders and even the ultrasecretive cryptographers who had broken the Japanese naval codes. His 800-page report to Stimson remained classified for years. Clausen revealed that Kimmel and Short on December 3, 1941, had had a sharp argument over sending Army troops to reinforce Wake Island (then held by U.S. Marines), and as a result of the debate over which service would then command at Wake, neither commander attempted to talk with the other and they did not talk or meet until after the Japanese attack. Kimmel did not inform Short that U.S. naval intelligence had lost track of the Japanese carrier force—then en route to Hawaii—and failed to use naval aircraft as ordered to carry out reconnaissance missions. Short himself failed to carry out orders from Washington to collaborate on the reconnaissance flights and closely cooperate with the Navy, and he failed to aggressively use new radar systems in hand to extend the surveillance umbrella out to the 130-mile range of the radar. Clausen in his 1992 account concluded, "So much for command by mutual cooperation."

3. See Geoffrey Regan, *Blue on Blue: A History of Friendly Fire* (New York: Avon Books, 1995) for a comprehensive history of fratricidal incidents in war.
4. See Richard A. Gabriel, *Military Incompetence: Why the U.S. Military Doesn't Win* (New York: Hill and Wang, 1985), pp. 85–116. Also see Colonel Charlie Beckwith with Donald Knox, *Delta Force* (New York: Dell Publishing, 1983), pp. 268–69: During a Senate hearing after the Iran rescue mission failure, Beckwith was asked by Senator Sam Nunn, a Democrat from Georgia, why the operation had failed. Beckwith's reply centered on the lack of coordination and interservice teamwork that preceded the actual raid. "Senator," Beckwith told Nunn, "I've known the answer to your second question since I was a captain. What do we need to do in the future? Sir, let me answer you this way. . . . If Coach Bear Bryant at the University of Alabama put his quarterback in Virginia, his backfield in North Carolina, his offensive line in Georgia, and his defense in Texas, and then got Delta Airlines to pick them up and fly them to Birmingham on game day, he wouldn't have his winning record. . . . He has a team. In Iran we had an ad hoc affair. We went out, found bits and pieces, people and equipment, brought them together occasionally and asked them to perform a highly complex mission. The parts performed but they didn't necessarily perform as a team."
5. *Army Times*, November 5, 1984, p. 34.
6. "The Acquittal of Capt. Wang Is Unlikely to Calm the Storm," *Air Force Times*, July 3, 1995. Also "Pilots Use Faulty IFF Equipment—Disclosure May Shed Light on the Shootdown over Iraq," *Air Force Times*, April 10, 1995.
7. *Who Goes There: Friend or Foe?* a report by the Office of Technology Assessment, June 1993.
8. Ibid., pp. 26–27.
9. The cultural differences among the combat services are highlighted in a well-told joke about how to secure a building: "Ask the Army to secure a building and it will deploy a platoon of military police around the site. Ask the Navy to secure the building and a sailor will padlock the front door when he leaves for the night. Ask the Marines to secure a building and they will level it with artillery fire. And ask the Air Force to secure the site and they'll take a five-year rental with an option to buy." An even more humorous, actual incident in 1987 experienced by the coauthor demonstrates that cultural differences from service to service are profound: The coauthor had spent a number of years observing the U.S. Navy and without consciously realizing it had absorbed awareness about the sea service's traditions without realizing they were not universally held—one of them being that senior naval officers always ride in the back seat of an official car with the more junior members in the front seat. One day the coauthor visited a U.S. Air Force bomber base and took an orientation flight on a B-52H bomber that sustained a major in-flight emergency at the instant of takeoff. After circling the region for several hours to burn off sufficient amounts of fuel to land safely, the aircraft returned for an emergency landing that ended safely with the bomber shut down on the main runway due to the fact that it had lost hydraulic power in

the systems to steer the aircraft on the ground. When the aircraft crew and journalist emerged from the aircraft into dazzling sunlight, the journalist could see the silhouette of an official Air Force sedan with the wing commander's logo parked several yards away. The journalist (because of his experience covering admirals) walked over to the car and automatically peered into the back seat looking for the wing commander—and as his eyes adjusted to the light, found himself gazing upon two second lieutenants sitting there. "I'm up here, you idiot," the wing commander growled from his place behind the steering wheel. Subsequent discussions in the intellectually rigorous confines of Happy Hour at the officer's club—the safe return of the crippled bomber called for a celebration of sorts—revealed the military roots of this cultural dichotomy (where to sit in cars): Navy admirals always sat in the back of the launch or cutter while the lower ranks did the rowing; Air Force leaders disdained the aircraft's passenger compartment for the glory of the cockpit and the pleasure of manipulating the flight controls. The journalist's suggestion that the Pentagon could save millions of dollars in transportation funds and extra drivers' salaries by having the leadership of the Air Force always drive the leadership of the Navy to meetings, was not adopted in any subsequent Pentagon budget.

10. There are innumerable examples of this but the two that stick in my mind are, first, the grim competition between the Strategic Air Command and the Navy that led to the "Revolt of the Admirals" after then Secretary of Defense Louis Johnson yielded to Air Force pressure and canceled the planned supercarrier USS *United States* in favor of the B-36 bomber in 1949; and eight years later, when Navy opposition to a rival Army intermediate-range missile prompted the Pentagon to *order* the Army not to attempt launching a space satellite so that the Navy's Vanguard missile could do so. Vanguard blew up on the launching pad, the Soviets launched their satellite *Sputnik* into space before the United States, and only then did the Pentagon allow an Army Jupiter-C launch—which was successful—with the first American satellite.
11. There were several dozen Army aircraft available for patroling the Atlantic coast area immediately after Pearl Harbor, but they were largely ineffective for attacking submarines and their range and navigational capabilities permitted only daylight, fair-weather searches. See Michael Gannon, *Operation Drumbeat* (New York: Harper and Row, 1990), pp. 242–69 and 298–99. Also, Norman Polmar and Thomas B. Allen, *World War II: America at War* (New York: Random House, 1991), p. 73. Once Army patrol planes were finally allowed to operate along the East and Gulf coasts in late spring of 1942, they made a strong contribution to the U.S. effort to suppress U-boat attacks, and one Army plane single-handedly sank the U-701 off Cape Hatteras on July 7, 1942.
12. The history of the crucial Battle of Leyte Gulf in 1944, when MacArthur first landed in the Philippines, offers a cautionary tale about the status quo of interservice rivalry during World War II. Nimitz's carrier commander, Vice Admiral William Halsey, ordered his carrier forces north and abandoned their role to protect

MacArthur's landing on Leyte when word came that the last Japanese aircraft carriers were within striking range. As a result, when Japanese Admiral Kurita steamed through the San Bernardino Straits toward Leyte Gulf, only a handful of ships manned by Navy reservists stood in the way of Kurita's ability to destroy MacArthur's landing force. What Halsey didn't know was that the Japanese carriers were devoid of experienced pilots and constituted a desperation feint to draw the more powerful U.S. Navy carriers away from the Leyte Gulf beachheads to clear the way for Kurita. As a result of the U.S. Navy's desire to annihilate the Japanese Navy at all costs, Halsey had rendered MacArthur's ground invasion force vulnerable to destruction. Only the individual bravery of the crews of several small U.S. Navy "jeep" carriers and their escorts thwarted Kurita and saved the day. Reflecting fifty-five years after this opaque example of interservice rivalry and miscommunication, I recall the words of Mark Twain: "God looks after children, drunks and the United States of America."

13. William W. Epley, *Roles and Missions of the United States Army: Basic Documents with Annotations and Bibliography* (Washington: U.S. Army Center of Military History, 1991).
14. Navy and Marine Corps leaders who merit the lion's share of the credit for transforming the Navy–Marine Corps strategic priorities at this key juncture include Admiral Frank Kelso, then chief of naval operations; General Carl E. Mundy, commandant of the Marine Corps; Vice Admiral Leighton "Snuffy" Smith, deputy chief of naval operations for plans, policy and operations, coordinator of the development of "From the Sea" (who would culminate his distinguished naval career as commander, Allied Forces Southern Europe during NATO's intervention in Bosnia in 1995); Brigadier General Thomas L. Wilkerson, then Marine Corps assistant chief of staff for plans, policies and operations; Vice Admiral Jerry Tuttle, who as director of naval space and electronic warfare, led the service's push for a communications revolution. "From the Sea" was a construct jointly prepared by General Chuck Krulak, myself, and Admiral Leighton Smith, with Chief of Naval Operations Admiral Frank Kelso's strong support and the new secretary's personal involvement. It was a very significant departure for a blue water Navy built during the fifty years of the Cold War. The paper itemized a major restructuring of Navy and Marine Corps roles and missions to include a new emphasis on expeditionary forces. The new strategy envisioned a more closely integrated Navy–Marine Corps team designed to operate overseas in response to humanitarian disasters, political crises short of war, minor military conflicts, and major theater wars. The heart of the transformation centered on the Navy's realization—and adaptation toward—the reality that conflict in the late twentieth century and beyond would most likely occur in shallow water environments such as the Persian Gulf or near coastal areas in trouble spots ranging from Africa to southern Europe.
15. I was greatly aided in this endeavor by a team of associates, including Rear Admirals Riley Mixon, Tom Ryan, and Dave Oliver, as well as my former chief of staff, Captain Joe Mobley (who would go on to become a vice admiral).

16. I went on to a new life as a corporate executive in San Diego. Several months later I was giving a speech at the opening of the San Diego Padres' season when someone handed me a note that Mike Boorda was dead of a self-inflicted gunshot wound. Boorda had learned that *Newsweek* magazine was preparing an article accusing him of improperly wearing "combat V" devices on two decorations he had received during service in Vietnam. After Boorda's death, retired Admiral Elmo M. Zumwalt, who had served as commander of U.S. Naval Forces in Vietnam before becoming chief of naval operations himself in 1970, examined Boorda's service record and the complex regulations for awarding "combat V" devices and concluded Boorda indeed had been qualified for the decorations.
17. The concept of joining floating components into an artificial island is not new. During World War II the U.S. Navy welded several logistics ships together to make a mobile maintenance base, and the British built several artificial islands in the English Channel to serve as air defense gun platforms. The Pentagon has also studied building a floating (but anchored) platform of several hundred acres in size to replace a helicopter base on Okinawa, although this idea apparently sparked enough opposition in Japan to scuttle the plan. See Owens, *High Seas* (Annapolis, Md.: Naval Institute Press, 1995), pp. 162–64.
18. "The Coming Defense Train Wreck," Center for Strategic and International Studies, Washington, D.C., March 23, 1999.

CHAPTER 5

1. Private communication with the author.
2. "Will We Heed Lessons of War in Kosovo?" by Michael G. Vickers (director, strategic studies at the Center for Strategic and Budgetary Assessments in Washington, D.C.), *USA Today*, June 16, 1999.
3. Walter Pincus, "CIA Predicts More Kosovo Bloodshed," *The Washington Post*, February 3, 1999, p. 14.
4. Steven Erlanger, "Kosovo Negotiators Will Look to Impose a Quick Settlement," *The New York Times*, February 4, 1999.
5. Thomas W. Lippman, "U.S. Miscalculations Traced to Albright," *The Washington Post*, April 7, 1999, p. 1.
6. Quoted in "NATO Has Ground Plan for Kosovo," by Joyce Howard Price, *The Washington Times*, April 12, 1999, p. 1.
7. A British defense official said in late April it would require a minimum of 30,000 ground troops to enter a "nonpermissive" environment in Kosovo, and later accounts upped the estimate to nearly 150,000. See "Clinton Joins Allies on Ground Troops—NATO to Weigh Conditions of Kosovo Mission," by William Drozdiak and Thomas W. Lippman, *The Washington Post*, April 23, 1999, p. 1; also Eric Schmitt, "War Stretches Pentagon's Resources," *The New York Times*, May 2, 1999.

8. "So Is It Yes or No?" Op-ed column by Richard N. Haass, *The New York Times*, April 13, 1999.
9. Molly Moore and Bradley Graham, "NATO Plans for Peace, Not Ground Invasion—Refugees' Return Is Alliance Focus," *The Washington Post*, May 17, 1999, p. 1.
10. Clinton interview with *60 Minutes II*, March 31, 1999.
11. "Viable Ground Option Would Take Months," news analysis by Steven Komarow, *USA Today*, April 19, 1999, p. 3.
12. Dr. Michael Evans, "Dark Victory," *U.S. Naval Institute Proceedings*, September 1999, pp. 33–37.
13. Blaine Harden and John Broder, "Clinton Tries to Win the War and Keep the U.S. Voters Content," *The New York Times*, May 22, 1999, p. 1.
14. John A. Tirpak, "The First Six Weeks," *Air Force* (magazine), June 1999, p. 29.
15. General Richard Hawley, comments to the Defense Writers Group, Washington, D.C., April 29, 1999, reprinted in *Air Force* (magazine), June 1999, p. 47.
16. William Drozdiak, "War Showed U.S.–Allied Inequality," *The Washington Post*, June 28, 1999, p. 1.
17. Stephen J. Glain, "NATO Is Expected to Consider Developing Surveillance System," *The Wall Street Journal*, May 19, 1999.
18. General Hawley comments to the Defense Writers Group, Washington, D.C., April 29, 1999, reprinted in *Air Force* (magazine), June 1999, p. 50.
19. See Rowan Scarborough, "Smaller U.S. Military Is Spread Thin," *The Washington Times*, March 31, 1999, p. 1; Rick Maze, "Shrinking Military Ranks Concerns U.S. Leaders," *Army Times* (newspaper), April 4, 1999; Eric Schmitt, "War Stretches Pentagon's Resources," *The New York Times*, May 2, 1999.
20. Erin Q. Winograd, "Shinseki Hints at Restructuring, Aggressive Changes for the Army," *Inside the Army*, June 28, 1999, p. 1.
21. See Rick Atkinson, *Crusade—The Untold Story of the Persian Gulf War* (Boston: Houghton Mifflin, 1993), pp. 17–19 and 31–33. The commander of the Apache battalion from the 101st Air Assault Division who led the first air strike of Operation Desert Storm in 1991 was Lieutenant Colonel Richard Cody, who eight years later as a brigadier general commanded Task Force Hawk, the Apache unit dispatched to the border of Kosovo.
22. See *Army Magazine*, October 1998, pp. 240 and 245.
23. See Paul Richter, "Critics See Idle Gunships as Sign of U.S. Hesitation," *The Los Angeles Times*, April 10, 1999, p. 1.
24. Ibid.
25. See David Atkinson and Hunter Keeter, "Apache Role in Kosovo Illustrates Cracks in Joint Doctrine," *Defense Daily*, May 26, 1999, p. 6.

CHAPTER 6

1. The Pentagon's decision to centralize the electronic warfare mission under the Navy's control by retiring Air Force EF-111 Raven and F-4G "Wild Weasel" aircraft and reassigning Air Force flight crews to the Navy EA-6B Prowler wing at Whidbey Island Naval Air Station, Washington, in 1995, set a precedent for this type of cross-service integration. Four Prowler squadrons designated to support Air Force intervention wings are jointly manned today by Navy, Marine Corps, and Air Force flight crews and support personnel.
2. Business Executives for National Security, "Tooth to Tail Commission," Update Report 33, May 25, 1999.
3. Robert Heinlein, *Starship Troopers* (New York: G. P. Putnam's Sons, 1959).
4. *Annual Report to the White House and the Congress*, May 1998, message from Secretary of Defense William A. Cohen.
5. Quoted in "Pushing Toward a Brave New World: Experiments . . . Are Defining a New Era That Could Turn the Military Upside Down," by Jon R. Anderson, *Navy Times*, January 5, 1998.
6. Quoted in "Air Force Stands Up Effort to Improve Urban War Concepts," by David Atkinson, *Defense Daily*, March 18, 1999, p. 2.
7. *Annual Report*, May 1998, Chapter 13, p. 5.
8. *Army Focus 1994: Force XXI—America's Army in the Twenty-first Century*, Department of the Army, p. 2.
9. The Army schedule as of last year called for the 4th Infantry Division, with two brigades at Fort Hood, Texas, and a third brigade at Fort Carson, Colorado, to be fully "digitized" by 2000; the III Corps headquarters and 1st Cavalry Division by 2004; followed by digitization of the "light forces" of the XVIII Airborne Corps, including the 82nd Airborne Division, 101st Air Assault Division, 10th Mountain Division, and 3rd Infantry Division; and finally, the 2nd Infantry Division, 1st Infantry Division, and 1st Armored Division between 2004 and 2009.
10. The M270 Multiple-Launch Rocket System, first used in Operation Desert Storm, fires either a number of long-range free-flight rockets armed with powerful submunitions, or the Army Tactical Missile System—a conventionally armed ballistic missile capable of delivering "brilliant" cluster warheads that seek out individual moving vehicle targets. The rockets themselves are being upgraded for pinpoint accuracy.
11. *Final Report*, Division XXI Advanced Warfighting Experiment, U.S. Army Training and Doctrine Command, p. ES-1.
12. See *Jane's Defence Weekly*, April 28, 1999, interview with General Dennis Reimer, Army chief of staff.
13. Douglas A. Macgregor, *Breaking the Phalanx: A New Design for Landpower in the Twenty-first Century* (Westport, Conn.: Praeger, in cooperation with the Center for Strategic and International Studies, 1997), p. 49.

14. Macgregor was recently portrayed as "the most prominent reformer on active duty" in the Army, whose book, *Breaking the Phalanx*, had caused such a furor within the service's leadership that he was "given a refuge" by NATO commander General Wesley Clark, who had personally requested his assignment on his staff. See "With Army Suffering Identity Crisis, Choice of Chief Gains New Importance," by Thomas Ricks, *The Wall Street Journal*, March 3, 1999.
15. Macgregor, *Breaking the Phalanx*, p. 3.
16. "Another View of Information Warfare," in *The Information Revolution and National Security*, ed. by Stuart J. D. Schwartzstein (Washington, D.C.: Center for Strategic and International Studies, 1996).
17. The "Army After Next" project envisions U.S.-based forces capable of rapid deployment worldwide, and once in the crisis area, capable of tactical movement of hundreds of miles a day using advanced tactical airlift systems. See *Knowledge and Speed: The Annual Report on the Army After Next*, the Annual Report on the "Army After Next" Project, Department of the Army, July 1997, Appendix C.
18. Cited in *Knowledge and Speed*, p. 15.
19. Ibid.
20. See Sean Naylor, "Strike Force Concept Could Spread," *Army Times*, May 11, 1998.
21. *Joint Vision 2010*, p. 32.
22. Trevor Dupuy, "Theory of Combat," *Brassey's Encyclopedia of Military History and Biography* (Washington, D.C. and London: Brassey's, 1994), p. 975.
23. See United States House of Representatives, *Report of the Select Committee on U.S. National Security and Military/Commercial Concerns with the People's Republic of China*, Vols. 1–3.
24. Su Kuoshan, "Road of Hope—Reviewing the Accomplishments of the '963' Program on the 10th Anniversary of its Implementation," in *Jiefangjun Bao*, April 5, 1996, trans. by the Foreign Broadcast Information Service Daily Report, May 8, 1996 (FBIS-CHI-96-089).
25. *Report of the Select Committee on U.S. National Security and Military/Commercial Concerns with the People's Republic of China*, Vol. 1, pp. 10–13.
26. Ibid., pp. 13–14.
27. Ibid., pp. 18–19.
28. Richard D. Fisher Jr., "China Rockets into Military Space," *Asian Wall Street Journal*, December 29, 1998.
29. Jeremiah panel findings quoted in Maggie Farley, "China's Military Power Lags U.S. Despite Spy Fears," *The Los Angeles Times*, June 1, 1999, p. 1.
30. The current unified command plan assigns some unique responsibilities to promote joint operational concepts to the U.S. Atlantic Command (or USACOM). Like all U.S. regional commands, USACOM has command responsibility for all U.S. military forces operating in or transiting a particular geographic area, in this case the Atlantic Ocean. But USACOM has also been assigned the responsibility for preparing forces in the United States for their assignments overseas. This responsibility has been and remains controversial. Its boundaries have not been defined with pre-

cision, and the first USACOM commander in chief, Admiral Paul David Miller, took a very activist view of the responsibility. In his view, it encompassed the authority to define the character and mix of the forces needed by other regional commands. The other unified commanders differed markedly in their view, arguing, at times quite vociferously, that Miller's job was simply to help get the forces they wanted to their destinations, not to define what forces they wanted. Accordingly, the role of USACOM in "preparing" forces has been treated gingerly since that early confrontation. I believe Miller was generally correct and far in advance of his time.

31. See Macgregor, *Breaking the Phalanx*, pp. 71–74.
32. *New York Times*, December 14, 1898.
33. *The Manila Times*, July 4, 1902.

INDEX

INDEX